A NATURALIST'S GUIDE TO THE

PRIMATES OF SOUTHEAST ASIA

East Asia and the Indian Sub-continent

A NATURALIST'S GUIDE TO THE

PRIMATES OF SOUTHEAST ASIA

East Asia and the Indian Sub-continent

Chris R. Shepherd and Loretta Ann Shepherd

JOHN BEAUFOY PUBLISHING

First published in the United Kingdom in 2017 by John Beaufoy Publishing Ltd
11 Blenheim Court, 316 Woodstock Road, Oxford OX2 7NS, England
www.johnbeaufoy.com

Photo Credits
Front cover: *main* Bornean Slow Loris © James Eaton/Birdtour Asia, *bottom left* Long-tailed Macaque © Chris R. Shepherd, *bottom centre* Bornean Orangutan © James Eaton/Birdtour Asia, *bottom right* Agile Gibbon © Stephen Hogg. **Back cover:** Mitred Langur © James Eaton/Birdtour Asia. **Title page:** Southern Yellow-cheeked Crested Gibbon © Marina Kenyon. **Contents page:** Siau Island Tarsier © Geoff Deehan

Main descriptions: photos are denoted by a page number followed by t (top), b (bottom), l (left) or r (right).

Andie Ang 67b, 70b; **Nick Baker/EcologyAsia.com** 50l, 50r, 65t, 133; **Hannah Barlow** 40l, 40r; **Keith Barnes** 135; **Udayan Borthakur/Aaranyak** 129t, 129b; **Alex Cearns/Houndstooth Studio** 7r, 12t, 12b, 141; **Zhao Chao** 151l, 151r, 152l, 152r; **Dilip Chetry** 61; **Ng Bee Choo** 125; **Camille Coudrat** 86t, 86b; **Shashank Dalvi** 10t, 10b, 88, 95, 110r, 127t, 127b, 128; **Nabajit Das** 29t; **Geoffrey Davison** 100t; **Gehan de Silva Wijeyeratne** 8, 9b, 93, 122t, 122b, 123; **Geoff Deehan** 21; **James Eaton/Birdtour Asia** 6r, 16r, 19, 25, 35b, 44, 45, 46, 54l, 54r, 62t, 62b, 84, 85t, 85b, 89t, 89b, 90, 91, 98, 103, 109, 119, 126t, 126b, 149, 164, 165, 166; **Gabriella Fredriksson** 56, 57t, 57b; **Sharon Gursky** 22; **Stephen Hogg/Wildtrack Photography** 142t, 143, 145, 163; **Jyrki Hokkanen** 148; **Jeremy Holden/FFI** 102; **Jason Hon** 53; **Rob Hutchinson** 121; **Milan Janda** 60; **Xu Jianming** 78t, **Kahoru Kanari** 134; **Niu Kefeng** 99b; **Marina Kenyon** 156t, 159, 160t, 160b, 161t, 161b; 78b, 79; **Martjan Lammertink** 52t, 52b; **Ch'ien C. Lee/Wild Borneo Photography** 33, 37, 41, 120t; **Mark Lopez** 36t, 36b; **Joey Markx** 26l; **Nick Marx/Wildlife Alliance** 71, 140t, 140b; **Anuar McAfee** 142b; **Jeanne McKay** 16l; **Stefan Merker** 23t, 23b; **Alice Miles/Danau Girang Field Centre** 58t, 58b, 59t, 59b; **Andrea Molyneaux** 42, 43; **Richard Moore** 30t, 30b, 31, 35t; **Tilo Nadler/Endangered Primate Rescue Center** 64t, 64b, 76t, 76b, 77t, 77b, 81t, 81b, 82t, 82b, 154l, 154r, 155, 156b, 157, 158t, 158b; **Nature Picture Library/Xi Zhinong** 96, 99t; **Mike Nelson** 18; **Roberto Nistri** 65b; **Kara Norby** 146, 147; **Fan Pengfei** 138t, 138b; **Tim Plowden** 80, 83, 105b; **Christin Richter** 49t; **Bernat Ripoll/Borneo Nature Foundation** 144; **Noel Rowe** 26l, 117; **Fabian Schmidt** 118; **Jan Schmidt-Burbach** 107b; **Elke Schwierz** 74t, 74b; **Myron Shekelle** 17, 20, 24; **Chris R. Shepherd** 7l, 9t, 51l, 51r, 69, 72l, 107t, 108, 112, 131t, 131b, 132, 162; **Coke & Som Smith/www.cokesmithphototravel.com** 27, 28, 47, 63, 66, 73l, 87, 92, 94l, 94r, 97, 100b, 101, 106, 110l, 111, 115t, 115b, 120b, 124, 130, 137, 150, 153; **Ulrike Streicher** 38, 104; **Syahputra** 34; **Rob Tizard** 139; **Tontan Travel** 73r; **Larry Ulibarri** 105t; **Andrew Walmsley** 14t, 14b, 32t, 32b, 116; **Serge Wich** 39l, 39r; **Peter Widman** 68; **Christy Williams** 6l, 136l, 136r; **Roland Wirth** 113, 114; **Peter Yuen** 70t, 72r; **Thomas Ziegler** 48, 49b.

ISBN 978-1-909612-24-2

Dedication
For our daughters Raven Dhanya Shepherd and Robyn Lyla Shepherd, our favourite little monkeys

Edited by Krystyna Mayer

Designed by Gulmohur Press, New Delhi

Printed and bound in Malaysia by Times Offset (M) Sdn. Bhd.

Contents

Introduction

Primates are among the most conspicuous of the world's mammals, and 120 species occur in Asia. Most are diurnal and many are gregarious. In places where they face minimum threats from humans, they can be common and sometimes easy to see. Other primates are nocturnal and secretive by nature and require far more effort, skill and patience to find in their natural habitats.

Asia's non-human primates use a wide variety of habitats, from the hot and humid mangroves, through a vast variety of forests, up to high, snow-covered montane forests. Only in the highest and harshest of mountain environments and the driest of deserts are primates naturally absent. Many species have specialized habitat requirements – some langurs, for example, utilize rough and jagged limestone karst, where they forage, and find refuge from predators and caves to sleep in. Others, like Proboscis Monkeys, utilize humid lowland mangrove and riverine forests. Some species of macaque and snub-nosed monkey have adapted to life in cold and harsh, snowy environments.

Given primates' extremely varied habitats, it is of no surprise that their locomotion style varies greatly, from the swift leaps made by tarsiers between vertical trunks and branches, to the breakneck brachiating by gibbons through the canopy, and the slower and deliberate movements of stalking lorises.

All primates are arboreal, though many species, especially among the macaques, spend considerable time travelling and foraging on the ground. Some species live relatively solitary lives, meeting infrequently to mate or otherwise interact, while others live in largely monogamous pairs with their young, and still others live in groups that sometimes number more than a hundred.

Primates give birth to single young, although twins have been recorded in rare cases among some species. In some parts of the region with extreme seasonal variation, breeding takes place at select times of the year, but for the most part they breed year round.

Old World primates, those found in Asia, do not have the prehensile tail found in their New World counterparts. Many, such as gibbons, orangutans and lorises, have inconspicuous tails or altogether lack tails. Others, like tarsiers, langurs and macaques, have extremely long tails that are used for balance.

Sadly, an increasing number of Asia's primates are severely threatened by human-related

Gibbons brachiate through the forest canopy

Leaping rather than climbing to the ground

impacts – by hunting for local consumption and for the burgeoning illegal wildlife trade, and by habitat destruction, urban expansion, irresponsible agricultural practices and large-scale conversion of forests to monoculture plantations.

Conservation actions for Asia's primates have never been so urgently needed. Unfortunately, few people are aware of the primate diversity in Asia, or the threats these animals face. Fewer still are taking action to ensure that we do not lose any of these amazing species.

Watching primates and learning about them is fun, challenging and exciting, and we hope this book not only raises your appreciation for the primates of Asia and inspires you to learn more about these fascinating creatures, but also brings out the conservationist within you, encouraging you to get involved in protecting primates and the fragile world we share with them.

USING THIS BOOK

This book aims to introduce you to the 120 species of non-human primate found in Asia, with a peek into their world, providing photographs, physical descriptions, information on range and preferred habitats, and snippets on other unusual features. It is not intended to be a comprehensive field guide, or a work on taxonomy, but rather a simple reference for beginners, as well as a concise photographic guide useful while travelling in search of Asia's primates.

The descriptions are basic and do not cover all subspecies, but do provide enough information for identification. Measurements are provided for adult animals, largely following those in the *Handbook of the Mammals of the World (Vol. 3)*, edited by Russell A. Mittermeier, Anthony B. Rylands and Don E. Wilson, *The Pictorial Guide to Living*

Some live in large social groups

Few young, but strong maternal care

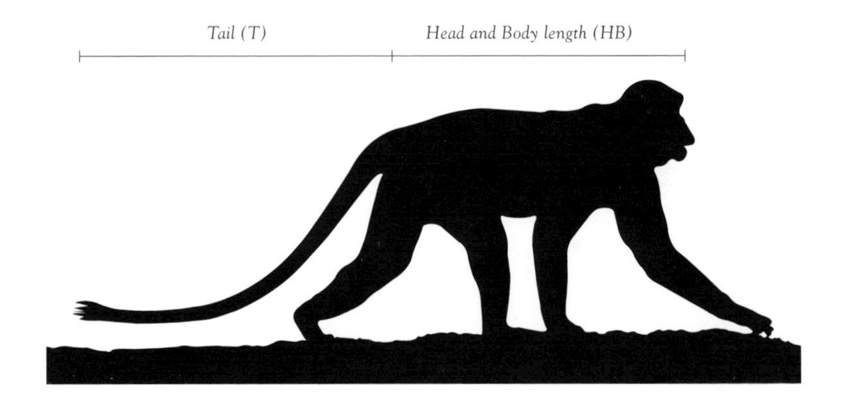

Primates, by Noel Rowe, and *A Field Guide to the Mammals of South-East Asia*, by Charles M. Francis. The caveat, however, is that for many species, insufficient work has been done to make definite size determinations, and the issue of size is further complicated by the greatly varying sizes and other morphological features through the range of many species. It is important to remember that sizes may vary greatly, and in some species there is great variety in both size and colour within a single species.

Each species account includes a list of countries the primate is native to, and in some cases gives the islands or areas to which a species may be endemic. However, areas to which species have been introduced outside their natural range are not included. A brief description of the known habitat or habitats is provided, along with a few notes on the diet and ecology of the species.

The checklist of all Asian primates is as complete and accurate as possible, but given the varying and often confusing taxonomic splitting and lumping together of primates, there may be some omissions or slightly outdated taxonomic revisions. It may also be that some of the species listed here are in fact subspecies and may be combined in the future.

The majority of the images used are of wild individuals photographed in their natural state, but a few are of captive individuals, used mainly because of their rarity or the unavailability of suitable wild images for identification.

NAMES AND TAXONOMY

Every species has a scientific (usually 'Latinized') name. This nomenclature is used to avoid confusion, as common names vary in different languages and locations. These scientific binominal (two-part) names are used universally. The first part, '*Pongo*' for example, denotes the genus, the second part, '*abeliii*', the actual species – *Pongo abelii* – the Sumatran Orangutan (see p. 165).

Common English names are also given to make it convenient for people who are

Play is important for bonding and learning survival skills

Some aggressively defend family and territory

unfamiliar with scientific names to recognize the species. However, common names can also be confusing, because there is often more than one name for a species (for example, Sundaic Silvered Langur, Silvered Leaf Monkey, Silvered Monkey and Silvery Lutung are all names for *Trachypithecus cristatus*). An attempt has been made to standardize the common names as far as possible, but note that other references may use other common names and therefore use of the scientific name is extremely important.

Taxonomy, especially among the langurs, lorises and tarsiers, has undergone multiple revisions, and is far from settled. More work, especially through the increasing use of DNA-based science, will undoubtedly help further sort out this fascinating puzzle. In addition to 'new' species being discovered in the lab, there may yet be species completely

A good pair of binoculars is useful when observing rare and shy species

unknown or undetected living in little-studied and remote parts of Asia; after all, the Myanmar Snub-nosed Monkey (see p. 101) was first described by science as recently as 2011, marking an incredibly exciting discovery in times when there is so much bad news in terms of increasingly threatened and endangered populations.

PRIMATE WATCHING

Looking for and observing primates is fun. Some basic tools needed are binoculars, identification guides, a notebook, a flashlight or headlamp to find nocturnal species, and a good camera as a bonus. Use red cellophane over your flashlight to reduce disturbance and discomfort to the primates (and other animals).

When recording observations of primates, it is important to note the habitat the animal is in, the altitude and other pertinent information. Protecting habitat is essential to the conservation of Asian mammals, and we currently do not know enough about the habitat requirements and preferences of most species. It is also important to record the behaviour you observe – what it is eating (or what is eating it!), how it is interacting with others of its kind and with other species, and any other interesting observations.

Some species are far easier to spot and watch than others. In areas where threats to primates are minimal, they are usually more common and easier to see, as are species that adapt well to human-altered habitats and can live in close proximity to people. However, many of the rarer species and those hunted are far more difficult to find, and even more difficult to observe. Often, the only glimpse you have of many species is of the back end of the animal as it leaps through the forest to safety.

Identifying primate species is not always easy, especially where clear views are not possible because they are hidden in dense foliage, or are fleeing through the forest canopy. It takes some practice and a lot of patience. Do not be discouraged.

Identification of primates under field conditions is made easier given that the different primate taxa have allopatric distributions and not every species occurs everywhere. In many parts of Asia, the typical primate community comprises one species of (slow or slender) loris, one, two or three species of macaque, and one or two colobines (often from different genera). In Southeast Asia we can expect to add one species of gibbon to this assemblage. Along the western and northern edges of the primate range in Asia we typically see a very impoverished primate community, with often just one species, whereas in the south-east, primates like orangutans and tarsiers increase the total number. Interestingly, in parts of mainland Southeast Asia we can see up to four species of macaque living in the same general forest area, and in parts of Borneo up to five colobines.

It is important to follow some basic principles and ethics when observing primates – keep noise levels low, wear mute-coloured clothes, move quietly and cautiously, and pay attention to the sounds and signs. Do not harass the animals, and be considerate to others who may also be observing them. *Never* feed the primates because this almost always leads to human-animal conflict, with wildlife ultimately paying the price.

Some primates, such as orangutans and a number of the macaques, have been studied extensively, while others are very poorly known. Increased knowledge, whether through

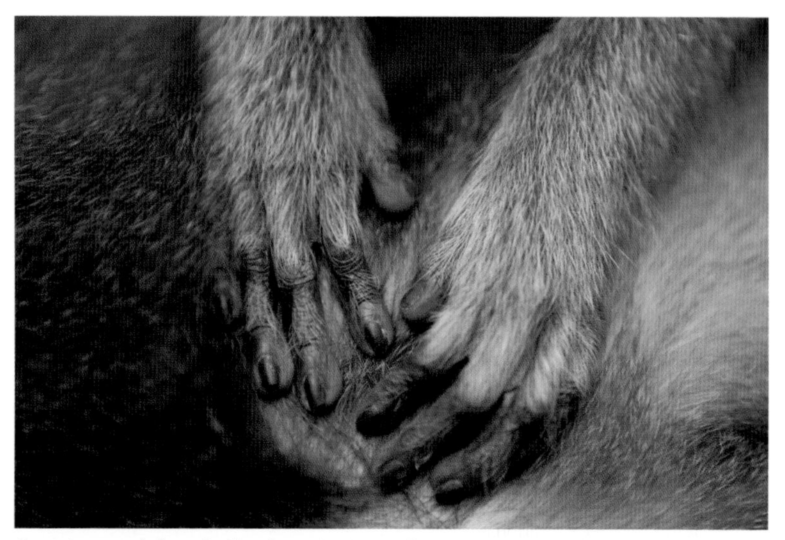

Grooming not only keeps fur clean but maintains social structure

Hands not unlike our own, used for climbing, manipulating objects and sometimes using tools

basic field observations, or resulting from intense research, is desperately needed in order to better understand the needs of each and to ultimately ensure that we do not lose any species from the wild. There is not a great deal of information on most of the region's mammals, so every observation, record and photograph taken is potentially an important contribution to the overall understanding and conservation of Asia's primates.

Primates in trouble

Across Asia, primates are in trouble. Human-related activities are directly or indirectly responsible for the threatened status of all these species, yet protecting them from over-exploitation and from the threats of habitat loss and degradation in most parts of Asia remains a very low priority.

HABITAT LOSS

Habitat use of many primate species may vary, sometimes seasonally, sometimes due to disturbances. Some species are generalists and use a wide variety of habitats or are able to adapt not only to a number of natural settings, but also to alterations made by humans. Others leave or simply perish once their habitat has been disturbed.

Habitat loss and fragmentation are serious threats; residential and agricultural expansion, especially by monoculture plantations such as those of palm oil and rubber, has replaced large areas of habitat crucial to many mammal species. Highways and roads cut through forests, not only limiting the movements of many species, but also increasing access for hunters. Logging activities degrade and destroy pristine habitat. There are primates, such as some macaques, which can adapt and thrive in secondary forests, degraded habitats, and agricultural and even urban areas. Most, however, cannot survive in the long term in highly disturbed habitats. Large-scale monoculture plantations in particular are a major threat to the continued survival of many species, destroying habitats, isolating populations and bringing some species into conflict with people. In almost all cases, the wildlife is the loser in such conflicts.

Steps must be taken to minimize the negative impact of such development. Key habitats and connecting corridors must be set aside. Long-term land-use planning, be it for plantations, roads, farming or other development, must take the conservation of wildlife and wild places into consideration. In Asia, where human population growth and expansion is taking place at a frightening pace, this is a major challenge.

HUNTING AND TRADE

Illegal and unsustainable hunting and trade in primates, both domestic and international, is a growing threat to a rapidly increasing number of species in Asia, especially in Southeast Asia. Wildlife has been harvested and traded throughout this region for thousands of years, yet never have the levels been as intense and as destructive as has been observed over the past few decades. Macaques, for instance, are trapped in extremely large

Illegal trade is a major threat to many of Asia's primates

numbers for biomedical purposes, and for local consumption in the form of traditional medicines and food. Other primates, such as lorises, are traded as pets, used as props in tourist spots, or killed and their parts used in traditional medicines.

Among all the threats wildlife faces, illegal trade is an extremely urgent issue that needs the highest levels of attention, as it has the greatest potential to do maximum harm in a short time.

HUMAN-WILDLIFE CONFLICT

A pressing problem in conservation is human-wildlife conflict – when people and wildlife share the same space and compete for natural resources. As the human population grows and encroaches into primate habitat, these animals are commonly killed, harassed or driven out of their habitats because they

Many primates are kept in cruel conditions as pets

are considered 'pests' that consume or damage human crops and other natural resources, or even cause harm to humans.

Primates that live near urban settlements are especially vulnerable victims as they become habituated to human presence and by easy access to human food (accidentally due to poor waste management, or deliberately through supplementary feeding). People who feed monkeys under the misguided belief that wild animals should be fed, or because they want interaction with wildlife, are unwittingly doing a great disservice to the animals. Monkeys used to being fed by humans many turn aggressive and pose a danger to people. The authorities may relocate some of these 'problem' animals, but the majority of them are culled. Primates living near villages, orchards and plantations converted from natural forests often raid crops, and they are frequently eradicated with poisons, traps and guns. There have also been some extremely brutal killings of orangutans in oil-palm plantations by workers, using machetes and sticks.

GLOSSARY

allopatric Occurring in different geographical areas.
anthropogenic Caused by humans.
axillary Near the armpit region.
brachiate To move by swinging arms from one hold to another.
colobine Members of the Colobinae family of primates.
commensal Two species living in close association but without affecting each other.
conspecific Of the same species.
crepuscular Active at dawn and dusk.
dipterocarp trees Tall hardwood tropical trees that can grow to exceptionally large sizes.
distal Away from the body, towards the end.
diurnal Active during the day.
dorsal Back surface part of the body.
endemic species Native species restricted to a particular geographic area.
frugivorous Existing on a main diet of fruits.
karst Landscape generally underlain by limestone or dolomite.
melanistic Dark pigmentation.
montane Mountainous area.
nocturnal Active during the night.
pantropical Throughout the tropics.
paranasal Near or adjacent to the nose.
pelage Coat of a mammal (such as fur, wool or hair).
prehensile Tip (mainly of tail, snout or lips) that can curl and grasp objects.
primary habitat/forest Old-growth forest that has never been logged.
secondary habitat/forest Forest that has been logged but has recovered.
split (taxonomy) Dividing further, to represent more than one group.
submontane Foothills of mountainous areas.
sympatric More than one species sharing the same geographic area.

TARSIIDAE (TARSIERS)

These nocturnal primates are strictly arboreal, extremely agile and move swiftly under the cover of darkness. They are distinct creatures because of their very small size and disproportionately large eyes and ears, which they use to locate their prey in the darkness. Tarsiers are named for their elongated tarsus (ankle bone), which enables them to jump vast distances, up to a staggering 45 times the head and body length. They have short forelimbs, and use them to climb vertically on vines and branches.

Tarsiers can swivel their heads 180 degrees, much like owls, as their eyes are fixed in their sockets. All tarsiers are carnivorous, feeding on large insects such as crickets, moths and cockroaches, and small vertebrates such as birds, snakes and bats.

There is much debate surrounding the taxonomy of tarsiers, and some authors include more species than are included here.

Spectral Tarsier ■ *Tarsius tarsier* HB 12–14cm, T 23–26cm

DESCRIPTION Overall grey pelage on the body and thighs – it is generally more grey than in the mainland Sulawesi tarsiers. This species has a conspicuous black paranasal spot, as well as buffy or white hairs on the sides of its upper lip. It has a sparse, short tail-tuft. **DISTRIBUTION** Indonesia (Selayar Island, off the tip off south-west peninsula of Sulawesi). **HABITAT AND HABITS** Lives in mostly degraded and secondary growth patches and farmland scrub on Selayar Island. The island is not mountainous, but it is found from sea level up to the highest hills there. Specific dietary information is lacking, but it probably feeds on insects and small vertebrates. **Notes** Selayar Island has already lost much of its forest habitat, and currently is a proposed site for a large oil refinery that would probably pose a severe threat to the continued survival of this species.

Makassar Tarsier ■ *Tarsius fuscus* HB 12cm, T 24–26cm

DESCRIPTION Overall rufous-brown colour above with creamy underparts, with a long, tufted tail that widens at the tip. There is a white patch behind each ear, and a black spot on both sides of the snout. **DISTRIBUTION** Indonesia (south-west peninsula of Sulawesi). **HABITAT AND HABITS** Found in primary and secondary tropical forests, thorn scrub, coastal mangrove forests and montane forests in the south-west peninsula of Sulawesi. Also occasionally occurs in plantations and gardens. This species has been discovered to sleep in a matrix of small holes and interconnected tunnels in karst hills. Like other tarsiers, it most probably eats insects and small vertebrates.

Dian's Tarsier ■ *Tarsius dentatus* HB 12–13cm, T 22–25cm

DESCRIPTION Greyish-buff colour, with a hairless tail except for some hair on the tip. Short white hairs flank its upper lip and the middle of its lower lip. **DISTRIBUTION** Indonesia. East section of the central core of Sulawesi to the tip of the eastern peninsula, with the northern boundary being the Isthmus of Palu between Marantale, Ampibabo and Tomini Bay, while the southern boundary from Lore Lindu National Park to the eastern coast of Sulawesi is unknown. The western boundary appears to extend to at least the Palu River and south as far as Gimpu. **HABITAT AND HABITS** Found in primary and disturbed secondary forests and mangroves. Lives in small family groups of 2–7 individuals, foraging for insects and small vertebrates by night, and sleeping in tree cavities and thick foliage by day.

Peleng Tarsier ■ *Tarsius pelengensis* HB 12–14cm, T 25–27cm

DESCRIPTION Mix of brown and grey with an especially long tail. Very similar to Dian's Tarsier (see opposite), and it is likely that these two species are closely related. **DISTRIBUTION** Indonesia (Peleng Island, off the coast of the eastern peninsula of Sulawesi, and possibly other islands of the Banggai Archipelago). **HABITAT AND HABITS** Little is known about this species, but based on what is known about other tarsiers, it is believed to inhabit lowland primary and secondary forests and mangroves. It is also likely to live in small, monogamous or polygamous groups of 2–6 individuals. Despite its overall similarity to Dian's Tarsier, there are distinct differences in the vocalizations of the two species.

Sangihe Tarsier ■ *Tarsius sangirensis* HB 15cm, T 29–31cm

DESCRIPTION Golden-brown fur on the back, and white on the stomach. Generally less woolly than other Sulawesi tarsiers. Very little fur on the tail and tarsi. Eyes are pale-chestnut in colour. **DISTRIBUTION** Indonesia (Sulawesi, on the Great Sangihe Island and possibly other islands in this chain). **HABITAT AND HABITS** Although this species has not been systematically studied in the wild, it has been observed to prefer primary forest, but is also found in secondary forest, coconut plantations and scrub habitats on the very small island of Sangihe. It lives in small groups of 2–6 individuals.

Siau Island Tarsier ■ *Tarsius tumpara* HB 15cm, T 26cm

DESCRIPTION Mottled brown with dark grey undercoat. Thick brown line borders the grey fur above and lateral to its eyes. Whitish hair around mouth and short, sparsely haired tail-tuft. Although it is quite like the Sangihe Tarsier (see opposite), it has a more greyish pelage. **DISTRIBUTION** Indonesia (Sulawesi, on the Siau Island and possibly other islands nearby). **HABITAT AND HABITS** This Critically Endangered tarsier is found only in lowland wet forest on the volcanic Siau Island, which lies off north Sulawesi, but it is possible that it also occurs on some very small islands near Siau. **Notes** The tiny island of Siau is dominated by an active volcano, which could prove a significant threat to the population depending on the magnitude of any eruption. The species is also threatened by local hunting for consumption.

Pygmy Tarsier ■ *Tarsius pumilus* HB 8–11cm, T 20–21cm

DESCRIPTION Grey to brown body and reddish face, and smaller ears compared with other tarsiers. This species weighs only 55g, and has a heavily haired tail. **DISTRIBUTION** Indonesia (south and central Sulawesi, known only from the Rano Rano and Latimojong Mountains). **HABITAT AND HABITS** Lives in montane and cloud forests up to 2,200m on Mount Rore Katimbu in the Lore Lindu National Park. Mainly eats insects. **Notes** The smallest of all tarsiers, this species was rediscovered in 2008, after 90 years of not being seen and thought to be extinct.

Lariang Tarsier ■ *Tarsius lariang* HB 12cm, T 24–25cm

DESCRIPTION Greyish-brown overall, and generally darker than the other Sulawesi tarsiers. Blackish tail that ends with a dark, pencil-like point. Clear, dark ring around eyes. **DISTRIBUTION** Indonesia (west-central Sulawesi in the Lariang River Basin). **HABITAT AND HABITS** Found in primary and secondary forests, mangroves, forest gardens and also other areas with some human disturbance. Similar in habits to the Spectral Tarsier (see p. 16), from which it has only recently been distinguished as a separate species. Eats mainly insects and some small vertebrates.

Wallace's Tarsier ■ *Tarsius wallacei* HB 11–12cm, T 24–27cm

DESCRIPTION Yellowish-brown, mottled fur with an off-white abdomen and a copper-coloured throat. Also has yellow to copper patches around the eyes, almost forming an eye-ring. Dark, long tail with thick, long tail-tuft. **DISTRIBUTION** Indonesia (north-west Sulawesi). **HABITAT AND HABITS** Found in primary, secondary and degraded forests, and known to occur in the Gunung Sojol Nature Reserve. Preys on insects. **Notes** Researchers named this recently described species after Alfred Russel Wallace, who was the first to discover the zoogeographic boundary now known as the Wallace Line. This species occurs immediately east of this line.

Philippine Tarsier ■ *Carlito syrichta* HB 8.5–16cm, T 14–28cm

DESCRIPTION Short but silky, wavy fur varying in colour from buff to greyish-brown, and dark brown upperparts, while the underparts are buff, greyish or slate. Long tail is hairless except for a few short hairs on the tip. **DISTRIBUTION** The Philippines (south-eastern Philippines, restricted to the greater Mindanao faunal region). **HABITAT AND HABITS** Lives in both primary and secondary forests, in groups of 2–6 individuals.

Western Tarsier ■ *Cephalopachus bancanus* HB 12–15cm, T 18–22cm

DESCRIPTION Varies in colour, with the subspecies on Borneo (*T. b. bancanus*) being more golden-orange and rusty-brown compared with others that are more ivory-yellow. Tuft of long hair at the end of the long, hairless tail. **DISTRIBUTION** Brunei, Indonesia (south Sumatra, Bangka, Belitung, Karimata, Serasen in the South Natuna Islands and Kalimantan) and Malaysia (Sabah and Sarawak). **HABITAT AND HABITS** Found in primary and secondary forests, and also forest edges and along coasts. Exclusive diet of animal prey, consuming mainly insects such as beetles, grasshoppers, butterflies, moths and ants. Also eats small vertebrates such as birds, bats and snakes. Infants weigh a mere 25g at birth, but grow rapidly, catching their own insect prey at four weeks of age. Forages low in trees, around 2m off the ground, but if frightened will move higher up.

LORISIDAE (LORISES)

These are among the most primitive of primates. They are small and nocturnal, recognizable by their stocky shapes, round faces with forwards-facing eyes and vestigial tails. They have slow, quadrupedal locomotion and do not leap, creeping in a slow. deliberate manner to avoid detection by predator and prey alike.

Slow lorises are the only venomous primates. The venom is secreted by glands on the insides of their elbows, and when they lick a gland the secretion is mixed with saliva, enabling them to inflict venomous bites.

Red Slender Loris ■ *Loris tardigradus* HB 20–21cm

DESCRIPTION Very slender form, with a long and pointed muzzle. Reddish-brown fur with or without a dorsal stripe, and yellowish-white ventral hair. Short fur covers legs, feet and forearms. **DISTRIBUTION** Sri Lanka (central and south-western part). **HABITAT AND HABITS** Found in wet lowlands, tropical rainforests, and swampy coastal, evergreen and wet-zone lowland forests. Diet consists mainly of insects, and lizards including geckos. Lives in small family groups of 3–6 individuals. Vocalizes extensively, with calling bouts of more than 350 calls an hour having been recorded. **Notes** This is Sri Lanka's smallest primate, weighing about 130g. Numbers are less than 2,500 mature individuals, and these are threatened by habitat loss, roadkill, and poaching to supply the pet and traditional medicine trade, as well as by superstitious killing.

Grey Slender Loris ■ *Loris lydekkerianus* HB 21–26cm

DESCRIPTION Greyish coat with yellow on the muzzle, eyelids and ears, and teardrop dark frames around the eyes. The highland subspecies is most heavily furred. **DISTRIBUTION** India and Sri Lanka (south and eastern India, and north and central Sri Lanka). **HABITAT AND HABITS** Found in a variety of habitats, up to 1,200m above sea level. Main diet consists of insects, including those that contain toxic chemicals such as ants and bombardier beetles. Also eats flowers, berries and gum in a much smaller proportion. Lives in groups of up to 11 individuals. **Notes** Males, not just the fathers but also brothers, play a major role in infant care, which is extremely rare among primates. At two months of age, the mothers park the infants and do not visit them throughout the night, but the males do. Slender lorises regularly give birth to twins.

Bengal Slow Loris ■ *Nycticebus bengalensis* HB 30–38cm

DESCRIPTION The largest of the lorises, with brownish-grey to orange-brown, woolly upperparts, a pale underside and a distinct dark dorsal stripe that ends on the back of the nape. **DISTRIBUTION** Bangladesh, Cambodia, China (southern Yunnan and possibly southern Guangxi), north-east India, Laos, Myanmar, Thailand and Vietnam. Possibly occurs in northern Peninsular Malaysia. **HABITAT AND HABITS** Found in evergreen, deciduous and degraded forests, preferring forest edges with canopy cover, from sea level to 2,400m. Mainly eats tree gum, nectar, fruits, insects and eggs. **Notes** This species is hunted in parts of its range for its body parts in traditional medicine, for food and live for pets.

Sunda Slow Loris ■ *Nycticebus coucang* HB 26–30cm

DESCRIPTION Fur colour varies from light grey-brown to reddish-brown, with a dark dorsal stripe that extends from the top of the head to the lower back. On the back of the neck this stripe branches into four lines connecting to the ears and eyes. Dark rings encircle the large eyes. **DISTRIBUTION** Indonesia (Sumatra and some offshore islands), Malaysia (Peninsular Malaysia including the islands of Penang, Langkawi and Tioman), Singapore and southern Thailand. **HABITAT AND HABITS** Found in primary and secondary lowland forests, as well as gardens and plantations. Diet consists of gum, and also some fruits, insects, birds' eggs and leaves.

Javan Slow Loris ■ *Nycticebus javanicus* HB 30–39cm

DESCRIPTION Yellowish-brown with a dark dorsal stripe. Head, shoulders and neck are paler. Prominent creamy-white diamond shape between the eyes, formed by a distinct stripe that starts at the top of the head and forks towards the eyes and ears, extending down to the cheeks. Dark fur on ears. **DISTRIBUTION** Indonesia (Java). **HABITAT AND HABITS** Nocturnal and arboreal, moving slowly between hanging vines and branches like other lorises. Found in both primary and secondary forests. Main diet comprises the gum of trees, as well as insects. Sleeps curled up, hidden in branches. **Notes** Like all species of slow loris in Indonesia, this one is severely threatened by capture for the illegal pet trade.

Javan Slow Loris

Philippine Slow Loris ■ *Nycticebus menagensis* HB 26–38cm

DESCRIPTION Very small in size. Pale golden to red fur, with almost no markings on the round head. Very short ears and large, round eyes. **DISTRIBUTION** Brunei, Indonesia (East Kalimantan), Malaysia (Sabah) and the southern Philippines (Bongao, Sangasanga and Tawi Tawi). **HABITAT AND HABITS** Found in both primary and secondary lowland forests, as well as gardens and plantations. Omnivorous, eating the gum from woody vegetation, as well insects and animal matter. Detailed information on its diet is lacking, but research is ongoing. **Notes** Until recently, it was considered a subspecies of the Sunda Slow Loris (see p. 30).

Bangka Slow Loris ■ *Nycticebus bancanus* HB 26cm

DESCRIPTION Reddish upperparts, with lighter brownish-grey underparts and complete reddish-tinged rings around the eyes. Lacks the 'frosting' look on the fur present in some other species of slow loris. Ill-defined dorsal stripe. **DISTRIBUTION** Indonesia (West and South Kalimantan, south of the Kapuas River and east to the Barito River, and on the island of Bangka off Sumatra). **HABITAT AND HABITS** Found in a wide variety of forest types, ranging from primary and tall secondary forests, to peat-swamp and coastal forests, and evergreen, deciduous and gallery forests. Feeds largely on gum, but also takes insects from trees. **Notes** Until recently this species was considered a subspecies of the Philippine Slow Loris (see p. 33). Very little is known of this species' habits in the wild.

Bornean Slow Loris ■ *Nycticebus borneanus* HB 26cm

DESCRIPTION Upperparts greyish-brown with a reddish tinge, and lighter greyish-brown underparts. Dark contrasting mask around the eyes, which does not extend further down the face than the cheeks. Patch on top of the head and a varied, often broken dorsal stripe.

DISTRIBUTION Endemic to Borneo. Indonesia (West, South and Central Kalimantan, south of the Kapuas River as far east as the Barito River).

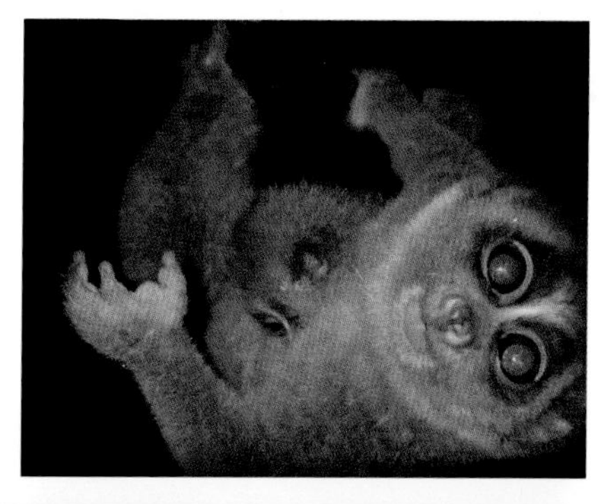

HABITAT AND HABITS Found in a variety of forest types, including both primary and secondary forests. Feeds largely on gum, but also takes insects from trees. Nocturnal and strictly arboreal. **Notes** This species was until recently considered a subspecies of the Philippine Slow Loris (see p. 33). Very little is known about it and more research is needed.

Kayan Slow Loris ■ *Nycticebus kayan* HB 27cm

DESCRIPTION The dark rings that encircle the eyes are rounded or pointed, not diffused at the edges; the bottom of each ring stretches low to the cheekbone, sometimes to the jaw. Stripe between the eyes is sometimes bulb shaped. Compared with the Philippine Slow Loris (see p. 33), the face-mask of the Kayan Slow Loris has more contrast between its dark black-and-white features, and its fur is generally longer and fluffier. **DISTRIBUTION** Endemic to Borneo (central and northern), Indonesia (Kalimantan), and Malaysia (Sabah and Sarawak). **HABITAT AND HABITS** Main diet consists of insects, tree gum, nectar and fruits. **Notes** Recent research focusing on the face-masks of the Bornean and Philippine lorises led to the recognition of four species – *N. menagensis*, *N. bancanus*, *N. borneanus* and *N. kayan*, with only *N. kayan* being new to science. *N. bancanus* and *N. borneanus* were previously considered subspecies of *N. menagensis*. Some of the species on Borneo may occur sympatrically.

Pygmy Slow Loris ■ *Nycticebus pygmaeus* HB 21–29cm

DESCRIPTION Thick and short, woolly fur that is light brownish-grey to orange and reddish-brown, with silver frosting. Underparts are light grey and ears are relatively conspicuous. Dorsal stripe is faint or absent, and the large eyes are encircled by dark rings. Pygmy Slow Lorises go through seasonal changes, differing greatly in size, colouration and marking between the cold, dry and warm, wet seasons. **DISTRIBUTION** Cambodia (east of the Mekong River), Laos and Vietnam. Presence in China is unsure (it is uncertain if records from south-eastern Yunnan are of native or introduced animals). **HABITAT AND HABITS** Primary evergreen and semi-evergreen forests, limestone forest, and bamboo, secondary and even highly degraded habitats. Forages alone, eating mainly insects and gum, as well as other plant matter and small animal prey such as geckos and birds. **Notes** Pygmy Slow Lorises occur sympatrically with Bengal Slow Lorises (see p. 29).

CERCOPITHECIDAE (COLOBINES)

The colobine monkeys in Asia include the langurs, doucs, snub-nosed monkeys and Proboscis Monkey. Species within this group are among the most spectacular, beautiful and varied of the primate world, with wide arrays of colours and patterns. Some species have brightly coloured young that gain their adult coats as they grow. Colobines are diurnal and mostly arboreal.

Often referred to as leaf monkeys, langurs and other Asian colobines have special sacculated stomachs that allow for the fermentation of otherwise difficult to digest cellulose. As a result, leaves make up the majority of the diets of most species in this group. The stomachs of colobines are often visibly enlarged, and contain micro-organisms that break down the leaves. Specialized molars also help break down the tough leaves. While some species, like those of the *Presbytis* genus, can eat more fruits or seeds than others, leaves make up the bulk of their diets.

The taxonomy of the langurs remains something of a mystery, with species frequently being 'lumped' and 'split'. Much more research is needed to truly understand this varied group of species.

Thomas's Langur ▪ *Presbytis thomasi* HB 42–61cm, T 50–85cm

DESCRIPTION Langur with striking facial markings. Eyes surrounded by white and grey-blue. Grey crest flanked by two white stripes, black moustache and flesh-coloured muzzle. Grey upperparts and white underparts, and a very long, pale tail. Hands and feet are black. **DISTRIBUTION** Indonesia (Sumatra – endemic to the northern provinces of Aceh and North Sumatra). **HABITAT AND HABITS** Occurs in primary and secondary rainforests, and neighbouring rubber plantations. Feeds mainly on young leaves, fruits and flowers, and also on small animal matter. Lives in family groups, usually of several females and one adult male, or occasionally two. After reaching maturity, males normally join into small groups of males, or lead solitary lives. Mainly arboreal, but occasionally descends to the ground to forage. **Notes** There are good chances of observing Thomas's Langurs near the Bohorok Orangutan Centre in Bukit Lawang and in the Gunung Leuser National Park.

Black-crested Sumatran Langur ■ *Presbytis melalophos* HB 42–57cm, T 64–82cm

DESCRIPTION Varies in colouration, ranging from a grey to whitish to reddish-orange coat. Inner legs and arms are white, and hands and feet usually orangish, although they are sometimes black. Colouration of outer side of limbs is pale, and orangish. Facial skin is black, with straw-coloured cheeks and a light forehead. Pointed crest is black. Newborns are light grey with a dark dorsal stripe. **DISTRIBUTION** Indonesia (western Sumatra). **HABITAT AND HABITS** Found in primary and secondary lowland, hilly and montane rainforests, and disturbed and plantation forests, where it feeds at all levels of the canopy on leaves, fruits, flowers and other plant matter. **Notes** Threatened by habitat loss and capture for the illegal pet trade. This species is considered by some to be a subspecies of *P. melalophos*, as *P. m. melalophus*.

Black Sumatran Langur *Presbytis sumatrana* HB 33–39cm, T 64–84cm

DESCRIPTION Dark grey-brown to almost black upperparts and dorsal surface of tail, with black feet, hands and outer surfaces of the limbs. There is a sharp contrast with the white underparts, throat, tail underside, and inner surfaces of the limbs until the wrists and ankles. Grey to dark brown crest and mostly bluish facial skin, with dark lips and flesh colour around the mouth. **DISTRIBUTION** Indonesia (western Sumatra and Pulau Pini in the Batu Archipelago). **HABITAT AND HABITS** Lives in lowland, hill old growth and secondary forests. Although it has not been studied extensively in the wild, it most likely has very similar dietary and behavioural habits to other *Presbytis* species. **Notes** This species is considered by some to be a subspecies of Black-crested Sumatran Langur (see p. 40), as *P. m. sumatrana.*

Black-and-White Langur ■ *Presbytis bicolor* HB 33–38cm, T 64–84cm

DESCRIPTION Overall dark brown body with black hands, feet, outer surfaces of hind limbs and upperpart of tail, and white underside, throat, inner surfaces of limbs and underside of tail. Crest has a black median strip and tip, and a black fringe of hair over the forehead. Also has a black muzzle. **DISTRIBUTION** Indonesia (east-central Sumatra). **HABITAT AND HABITS** Found mainly in the highlands from the Inderagiri River in the north to the Batang Hari River in the south. Similar to other *Presbytis* species, with a main diet of unripe fruits, leaves, flowers and seeds. **Notes** This species is considered by some to be a subspecies of Black-crested Sumatran Langur (see p. 40), as *P. m. bicolor*.

Mitred Langur ■ *Presbytis mitrata* HB 42–57cm, T 62–82cm

DESCRIPTION Varied in colour, usually with a dark grey upper body, outer limbs and head, but tinged with hues ranging from brown to ashy-grey, to orangish to yellowish-grey. Underside is creamy-yellow to whitish, as are the undersides of the limbs. Hands and feet are grey and the tail is reddish with a grey underside. Forehead and cheek tufts are whitish and there is a dark crest. Bare facial skin is greyish. **DISTRIBUTION** Indonesia (south-east Sumatra). **HABITAT AND HABITS** Found in primary and secondary lowland forests, and occasionally in mature rubber plantations. Feeds largely on leaves, but also takes unripe fruits, flowers and seeds. **Notes** Loss of forests due to the expansion of oil-palm plantations is the greatest threat to this species. It is also threatened due to illegal hunting and capture for the illegal pet trade, which is very widespread in Indonesia. This species is considered by some to be a subspecies of Black-crested Sumatran Langur (see p. 40), as *P. m. mitrata.*

Mitred Langur

Javan Grizzled Langur ■ *Presbytis comata* HB 51–54cm, T 65–66cm

DESCRIPTION Grizzled (grey-white mix) coat with a black head and crest, and paler underparts. Slender build, and longer hair frames the broad grey face. **DISTRIBUTION** Indonesia (western and central Java). **HABITAT AND HABITS** Though in the past it was found across extensive lowlands and mountains, habitat destruction has restricted it to patches of forest in montane habitats. Mainly eats leaves, fruits, flowers and seeds. Lives in troops usually consisting of 7–8 individuals, and occasionally forms groups with the Javan Langur. **Notes** Remaining populations live in some protected areas in western Java, including Ujung Kulon, Halimun and Gede-Pangrango national parks. Found at up to 2,500m asl.

Pagai Langur ■ *Presbytis potenziani* HB 50cm, T 58cm

DESCRIPTION Upperparts and long tail are black, while underparts are reddish to reddish-brown. Fur in genital region is yellowish-white, and male scrotum is white. White ring around black face. Newborns are light grey with a dark dorsal stripe, but begin to darken in colour within days of birth.

DISTRIBUTION Indonesia (endemic to Sipora, North and South Pagai, Mentawai Islands, off the west coast of Sumatra).

HABITAT AND HABITS Found in primary and secondary rainforests, swamp forests and mangroves, where it feeds largely on leaves, but also consumes fruits, seeds and other plant matter. **Notes** While Mentawai langurs live in small groups, this is the only monogamous colobine. It often associates with other primates when feeding, with the exception of Kloss's Gibbon (see p. 149), which displaces it at feeding sites. Severely threatened by hunting for human consumption, and habitat loss due to logging and conversion of forests to oil-palm plantations.

Siberut Langur ■ *Presbytis siberu* HB 48–50cm, T 58–64cm

DESCRIPTION Small langur. All black with reddish undersides, including undersides of limbs, and with some whitish fur around the face and neck, and around the pubic area. Slight forwards-facing crest of black hair. Infants born with white fur that darkens after 3–4 months. **DISTRIBUTION** Indonesia (Siberut Island, in the Mentawai archipelago, off the west coast of Sumatra). **HABITAT AND HABITS** Found in primary and tall secondary forests, and swamp forests, where most of its time is spent in the middle to upper canopy. Feeds on leaves and perhaps other plant material. **Notes** Until recently considered a subspecies of the Pagai Langur (see opposite). Seriously threatened by hunting, and habitat loss primarily due to logging. Expansion of oil-palm plantations in the archipelago is a threat to all primates there.

Common Banded Langur ■ *Presbytis femoralis* HB 46–59cm, T 69–77cm

DESCRIPTION Dark-brown to blackish upperparts, and grey underparts with pale patches on the inner thighs. Grey crest with pale grey skin around the eyes. **DISTRIBUTION** Indonesia (east and central Sumatra), Malaysia (disconnected distribution in Peninsular Malaysia, in the far south and north-west), Myanmar (south), Singapore and Thailand (south, in peninsular Thailand). **HABITAT AND HABITS** Lives in a wide variety of habitats, from mixed mangrove to primary and secondary forests. Like most langurs, its main diet consists of young leaves and fruits. **Notes** This is the only species of langur in Singapore.

White-thighed Langur ■ *Presbytis siamensis* HB 43–69cm, T 68–84cm

DESCRIPTION Brown to greyish-brown upperparts and tops of the head and arms; black hands, feet and distal half of the tail; and pale grey underside of the body, arms and legs, including a large patch on the outer thighs. Facial skin dark grey to almost black; occasionally, skin around the eyes is paler. Looks very similar to the Common Banded Langur (see opposite), but that species lacks the pale outer thighs. **DISTRIBUTION** Indonesia (Sumatra and Bintan Island in the Riau Archipelago, and possibly Batam and Galang Islands), Malaysia (Peninsular Malaysia) and southernmost Thailand. **HABITAT AND HABITS** Found in lowland to hill forests, including disturbed forests and plantations. Diurnal and arboreal. **Notes** This species is easily seen on Fraser's Hill and nearby mountains in Peninsular Malaysia, where it overlaps with Dusky Langurs (see p. 72) and other primate species.

Natuna Islands Langur ▪ *Presbytis natunae* HB 43–61cm, T 61–84cm

DESCRIPTION Dark greyish-brown upperparts, with lower parts of the limbs and head darker. White underside that extends to the backs of the thighs, chest, chin and sometimes ankles and wrists, but not the underside of the tail. Predominantly white, bushy cheek whiskers, and large white eye-rings that do not connect. Some depigmentation on the face. **DISTRIBUTION** Indonesia (endemic to the island of Bunguran in the Natuna Islands, off the north-west coast of Borneo). **HABITAT AND HABITS** Found in a range of habitats, including primary and secondary rainforests, heath forest and rubber plantations. Surveys have shown, however, that it prefers primary forests. Found at altitudes of up to 650m – at all altitudes on the island. **Notes** This species has been considered a subspecies of White-thighed, Common Banded and Black-crested Sumatran Langurs (see pp. 51, 50 and 40), before finally being accepted as a full species. It is threatened largely by habitat loss, although the illegal pet trade may also pose a serious threat. Overall, very little is known about this species and a great deal more research is needed.

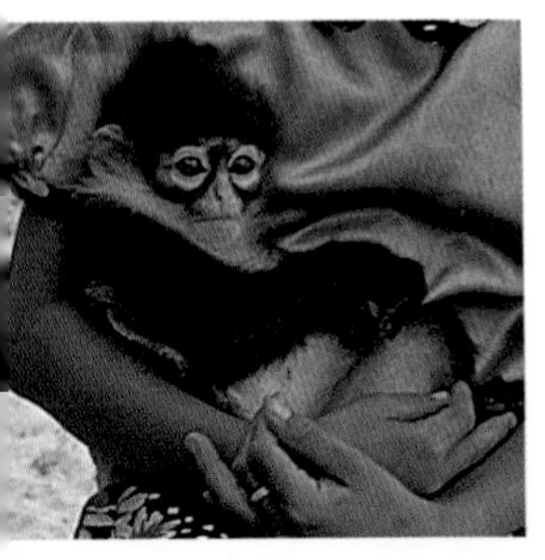

Bornean Banded Langur ■ *Presbytis chrysomelas* HB 46–59cm, T 70–186cm

DESCRIPTION Variable in colour, but mainly two separate colourations. One is usually dark brown, nearly black, with whitish underparts, inner limbs, underside of the tail, extending to the chin and cheeks. The other is brown with a pale or creamy-fawn underside, extending down the limbs and underside of the tail. Legs, flanks and crown are reddish. The young, which are light grey, have a striking cross-pattern on the back and shoulders, which is retained in some individuals to adulthood. **DISTRIBUTION** Endemic to Borneo. Brunei, Indonesia (west Kalimantan) and Malaysia (Sarawak and possibly Sabah). **HABITAT AND HABITS** Lives in lowland forests, including swamps and mangroves, feeding mainly on leaves, fruits and seeds, and lives in groups of 3–7 individuals. **Notes** Recent records come from five locations in Sarawak and West Kalimantan; it is Critically Endangered. Numbers have reduced by approximately 80 per cent over the last three decades, and there has been drastic loss of habitat. It now lives in less than 5 per cent of its historic range. This species is one of the rarest in the world and needs urgent conservation attention.

Hose's Langur ■ *Presbytis hosei* HB 48–56cm, T 65–84cm

DESCRIPTION This slim-built langur has a high forehead and prominent crest. There is grey fur on the back, the underparts are white, and the hands and feet are blackish. Facial skin is pink, with bold black markings in some subspecies. **DISTRIBUTION** Endemic to Borneo. Brunei (north-east), Indonesia (North Kalimantan) and Malaysia (west Sabah and Sarawak). **HABITAT AND HABITS** Lives in lowland to hill dipterocarp rainforests. Main diet comprises leaves, flowers, fruits, seeds, eggs and nestlings. Sympatric with the Maroon Langur (see opposite), sometimes associating closely with this species, which is unusual for Asian colobines. Mainly occupies mid-level of the forest canopy and descends to the ground to visit salt licks.

Maroon Langur ■ *Presbytis rubicunda* HB 44–58cm, T 67–80cm

DESCRIPTION Reddish-brown to golden-brown fur, and the face has a bluish tinge. Five subspecies exist, with slightly varying fur colouration. *P. r. chrysea*, which occurs in a small area near the Kinabatangan River in Sabah, is paler golden brown than the other subspecies, which are generally more reddish-brown. One of the subspecies occurring in Kalimantan, *P. r. rubicunda*, has blackish extremities to the limbs. The infants are white. **DISTRIBUTION** Endemic to Borneo. Occurs throughout most of Borneo, in Indonesia (Kalimantan and Karimata Island) and Malaysia (Sabah and Sarawak). Presence in Brunei uncertain. **HABITAT AND HABITS** Largely arboreal, preferring primary and secondary lowland to swamp forests, and also visiting gardens to feed. Consumes a large amount of seeds, and at least some populations are best described as granivorous. Also eats young leaves, fruits and flowers, but the diet varies according to availability of food sources. Lives in groups of up to 13 individuals, and males use loud calls to mark their territories.

Miller's Grizzled Langur ■ *Presbytis canicrus* HB 48–56cm, T 65–84cm

DESCRIPTION Grey upperparts with hairs tipped with white, giving it a grizzled appearance. Underside is light grey to whitish, and undersides of limbs are white. Hands and feet are black, as is the crest. Head is dark grey or blackish. Bare skin on the upper part of the face is dark reddish, turning pinkish at the lower parts of the face. **DISTRIBUTION** Endemic to Borneo. Indonesia (eastern Kalimantan). **HABITAT AND HABITS** Found in lowland and hilly dipterocarp forests, up to 1,600m. While largely arboreal, it does come to the ground, where it frequents mineral springs. Feeds largely on leaves, but also on fruits, seeds and even some animal material such as eggs and nestling birds. **Notes** Once considered a subspecies of Hose's Langur (see p. 55), Miller's Grizzled Langur is one of the rarest and least known species of primate in Asia. It has been hunted for the bezoar stones in its gut, which are believed by some to have medicinal properties. Conservation efforts are urgently needed to ensure the survival and recovery of this species.

Sabah Grizzled Langur ■ *Presbytis sabana* HB 48–56cm, T 65-84cm

DESCRIPTION This relatively small langur has a light grey body and tail, and a white belly and insides of the limbs. Hands and feet are black, as is the crest towards the rear of the head, and there are two black streaks on the sides of the head. Top off the head is grey. Bare skin on the face is orangish with black bands on the sides of the face. Lips are bluish. **DISTRIBUTION** Endemic to Borneo. Malaysia (throughout most of Sabah). **HABITAT AND HABITS** Found in lowland, riverine and montane forests, occasionally entering secondary forests and plantations. Feeds largely on leaves, though sometimes eats fruits, flowers, seeds and possibly the occasional eggs of birds. **Notes** Until recently considered a subspecies of Hose's Langur (see p. 55). Severely threatened by hunting for the bezoar stones sometimes found in its gut. These stones are considered by some to have medicinal value. Habitat loss is also a serious threat. May occur in Indonesian Kalimantan, though this remains unconfirmed.

White-fronted Langur ■ *Presbytis frontata* HB 53–60cm, T 63–74cm

DESCRIPTION Dark grey to greyish-brown upperparts, with darker arms, legs, hands and feet, and a lighter underside. Top and sides of head are dark, with a white spot on the head in a square or triangle shape, from which its name is derived. Chin and cheeks are a lighter grey. Tail is dark grey to brownish-grey. Crest on the head points slightly forwards. **DISTRIBUTION** Endemic to Borneo. Indonesia (east Kalimantan, and a few areas in west Kalimantan), and Malaysia (central Sarawak). **HABITAT AND HABITS** Patchy distribution. Found in lowland forests, as well as hilly and riverine forests. Feeds primarily on leaves, but also takes seeds, fruits and other plant material. **Notes** This is one of Borneo's least known primates, possibly because of its cryptic behaviour and small group sizes, hampering easy detection. Hunted for gallstones, and threatened by loss of habitat, especially due to conversion of forests to oil-palm plantations.

Gee's Golden Langur ■ *Trachypithecus geei* HB 50–75cm, T 70–100cm

DESCRIPTION Deep golden-orange to creamy-buff coat, with generally darker upperparts and sides, and paler underparts, though the fur changes seasonally; it is darker in winter and creamier in summer. Pale beard on dark, hairless face. Tail long and with a thick end. Infants are almost pure white. **DISTRIBUTION** Bhutan and India (north-east). **HABITAT AND HABITS** Lives in evergreen, dipterocarp, riverine, and moist deciduous and secondary forests. Main diet comprises leaves, fruits, buds and seeds. Prefers upper parts of trees, though it does venture to the ground to drink. **Notes** This species is found only in the foothills of Bhutan's Black Mountains, and a forest belt in western Assam in India.

Capped Langur ■ *Trachypithecus pileatus* HB 50–70cm, T 80–100cm

DESCRIPTION Distinct black cap and dark face; facial hair and underparts pale yellow to orange. Hairs on the crown are short, sticking straight up. Grey to pale brown body, and tail with black tip. Easy to distinguish from other langurs as none of the others have such a dark cap contrasting with a yellow-orange face. Juveniles up to the age of five months are creamy-white with a pink face. **DISTRIBUTION** Bangladesh (east), Bhutan, India (north-east) and Myanmar (north-west). **HABITAT AND HABITS** Found in evergreen, semi-evergreen, moist deciduous, bamboo and open woodlands. Diurnal and largely arboreal. Eats mainly leaves and also fruits, seeds and flowers. Lives in groups of multiple females and one male. **Notes** In some parts of its range, the Capped Langur is sympatric with Phayre's and Central Himalayan Sacred Langurs (see pp. 73 and 90).

Shortridge's Langur ■ *Trachypithecus shortridgei* HB 60–75cm, T 90–105cm

DESCRIPTION Overall silvery-grey coat, including the head and underparts; lighter legs and almost black hands, feet and tip of tail. Narrow, black-brown band and hair on top of head point upwards. Dark grey, hairless facial skin. Striking pale yellowish-orange eyes. Infants are bright orange. **DISTRIBUTION** China (south-west) and Myanmar (north-east). **HABITAT AND HABITS** Lives in evergreen and semi-evergreen broadleaved forests, and like other langurs feeds mainly on leaves and fruits.

Eastern Ebony Langur

■ *Trachypithecus auratus* HB 50–70cm, T 70–80cm

DESCRIPTION Throughout most of their range these langurs are black. In the easternmost part of Java, however, in addition to the black morph, red individuals occur alongside the black ones in the same troops, though the red morphs seem to be restricted to Blitar, Ijen and Pugeran. This langur has a small crest on its head, and the fur encircling its face points forwards, with prominent cheek-tufts. Its facial skin has a bluish tinge (or pinkish in red individuals). **DISTRIBUTION** Indonesia (east Java, Bali, Lombok, Sempu and Nusa Barung). **HABITAT AND HABITS** Found in mangrove and freshwater swamp forests, and lowland and montane forests up to 3,000–3,500m. Diet comprises mainly leaves, flowers, fruit seeds and unripe fruits. **Notes** Threatened, largely by habitat loss, but also by hunting for human consumption and sometimes for the illegal pet trade in rural areas, as well as in the large wildlife markets of Jakarta. The Ebony Langur (see p. 66) is sometimes treated as a subspecies of the Eastern Ebony Langur, but here it is treated as a full species.

Ebony Langur ■ *Trachypithecus mauritius* HB 44–65cm, T 61–87cm (based on Eastern Ebony Langur measurements)

DESCRIPTION Glossy black in colour, similar to the Eastern Ebony Langur (see p. 65), though lacking the 'frosted' fur and sometimes having a slight brownish tinge. Female has yellowish fur around the pubic area (like the Eastern Ebony Langur). Longish fur around the face curls forwards, and there are whitish hairs on the tips of the ears. Skin on the feet, hands and face is black. **DISTRIBUTION** Indonesia (western Java). **HABITAT AND HABITS** A wide variety of forests is used, including primary and secondary deciduous lowland and hilly forests, swamp forests, coastal forests and mangroves. Feeds largely on leaves, but takes other plant material such as fruits, seeds and flowers. **Notes** This species was until recently considered a subspecies of the Eastern Ebony Langur. It is threatened, largely by habitat loss, but also by hunting for human consumption and sometimes for the illegal pet trade, in rural areas as well as in the large wildlife markets of Jakarta.

Sundaic Silvered Langur ■ *Trachypithecus cristatus* HB 41–54cm, T 60–76cm

DESCRIPTION Overall dark grey with pale grey frosting, dark grey face and pointed crest. Infants are bright orange, but patches of grey appear until they adopt the full adult colouration. Long limbs and very long tail. **DISTRIBUTION** Brunei, Indonesia (Kalimantan, Sumatra, Bangka, Belitung and the Riau and Lingga archipelagos off eastern Sumatra, as well as the Natuna Islands), and Malaysia (Sabah and Sarawak). **HABITAT AND HABITS** Found in coastal, riverine and mangrove swamp forests. Diet comprises mainly young leaves, shoots, flowers, seeds and fruits, especially those of mangrove species. Lives in groups of 10–50 individuals. **Notes** The Selangor Silvered Langur (see p. 69) is often considered conspecific.

Selangor Silvered Langur ■ *Trachypithecus selangorensis* HB 46–54cm, T 68–82cm

DESCRIPTION Overall dark grey, including the hands, feet and face with no clear eye-rings. Long, straight whiskers and a high-pointed crest on top of the head. Infants are orange with white hands, feet and face. **DISTRIBUTION** Malaysia (Peninsular Malaysia – the west coast of Peninsular Malaysia, spanning the states of Kedah, Perak, Selangor, Negeri Sembilan, Melaka and Johor). **HABITAT AND HABITS** Found mainly in mangrove and riparian forests, and sometimes plantations. The main diet consists of leaves, followed by a much smaller proportion of fruits, seeds and flowers. **Notes** Can be seen easily at the Kuala Selangor Nature Park in the state of Selangor. Usually treated as conspecific with the Sundaic Silvered Langur (see p. 67).

Indochinese Silvered Langur ■ *Trachypithecus germaini* HB 49–57cm, T 72–84cm

DESCRIPTION Dark to medium grey upperparts, forearms and tail, with paler underparts and light grey patches on the limbs. Black hands and feet. General grizzled appearance because each hair has cream tips. Long, whisker-like white hairs frame the face. Short crest on top of the head and prominent brow ridges. Infants are bright orange. **DISTRIBUTION** Cambodia (west of the Mekong River), Laos (south), Myanmar (south), Thailand (southern Thailand, extending to Kanchamburi Province) and Vietnam (southernmost). **HABITAT AND HABITS** Lives in lowland forests, ranging from evergreen and semi-evergreen to mixed deciduous and riverine forests. **Notes** Widespread but extremely rare; especially threatened by hunting.

Annamese Langur ■ *Trachypithecus margarita* HB 55cm, T 72–84cm (based on Indochinese Silvered Langur because specific measurements for this species are not available)

DESCRIPTION Overall pale grey with very light underside and throat. Black forehead, forearms, hands and feet, and grey face with white eye-rings in many but not all individuals. Infants are golden-orange with white hands, feet and face. **DISTRIBUTION** Cambodia (east), Laos (south) and Vietnam (south-central). **HABITAT AND HABITS** Lives in lowland evergreen, semi-evergreen, mixed deciduous and riparian forests. There is no specific information on its feeding habits, but it is most probably mainly folivorous like other similar species. **Notes** Considered by some authorities as conspecific with Indochinese Silvered Langur (see opposite).

Dusky Langur ■ *Trachypithecus obscurus* HB 50–70cm, T 70–80cm

DESCRIPTION Distinct face with incomplete white rings around the eyes against dark grey facial skin, giving it the appearance of spectacles, hence its other common name, the Spectacled Leaf Monkey. Greyish-brown to dark grey upperparts with paler grey outer hind legs, tail and crest. Bare pink patches on the lips. Infants are light orange. **DISTRIBUTION** Thai-Malay Peninsula: Malaysia (Peninsular Malaysia including associated islands), Myanmar (south) and Thailand (south-western). **HABITAT AND HABITS** Inhabits primary and secondary forests, from lowlands to mountains, up to 1,800m. Prefers old-growth forests, but can be found in a wide range of disturbed habitats. Diet of mainly young leaves, shoots, flowers, seeds and fruits, especially those of mangrove species. Lives in groups of 10–50 individuals. **Notes** This beautiful langur can be easily seen on Fraser's Hill and other accessible destinations in Peninsular Malaysia.

Phayre's Langur ■ *Trachypithecus phayrei* HB 42–60cm, T 72–86cm

DESCRIPTION Overall, similar in appearance to the Dusky Langur (see opposite), with a dark grey body with paler limbs and tail. Dark, bare facial skin except for white eye-rings and some depigmentation patches on the lips and nasal area. Males can be distinguished from females by their ocular markings – in males, the white ocular rings around the eyes are parallel to the sides of the nose, resulting in a solid black strip, while the white ocular rings around the eyes in females are at an inwards-facing angle, resulting in a black triangular shape. Infants are straw coloured for their first few months. **DISTRIBUTION** Bangladesh (east), China (south-west – Yunnan Province), India (north-east) and Myanmar (north, west, east and south-east). **HABITAT AND HABITS** Lives in primary and secondary forests, including bamboo forest by hillsides and streams, eating mainly leaves and unripe fruits. Studies have shown that this species eats leaves from about 80 different plant species. As it eats some fruits, it may also play an important role as a seed disperser. **Notes** The Indochinese Grey Langur (see p. 74) is sometimes considered conspecific, but is here treated as a separate species.

Indochinese Grey Langur ■ *Trachypithecus crepusculus* HB 49–51cm, T 82–83cm

DESCRIPTION Light grey body and tail, with dark grey hands and feet. Silvery-grey belly and brown forehead, cheeks and hair around the face. Bare facial skin is dark grey, with full light grey rings around the eyes and some depigmentation patches on the lips. There is some geographic variation; in the north-east part of its distribution it is darker.

DISTRIBUTION China (south-west), Laos (north and central), Myanmar (south), Thailand (north) and Vietnam (north).
HABITAT AND HABITS Lives in primary and secondary evergreen and semi-evergreen, mixed moist deciduous forests, as well as bamboo-dominated habitats. Main diet comprises leaves, figs, bamboo shoots, seeds, flowers and gums.
Notes This species is sometimes considered a subspecies of Phayre's Langur (see p. 73).

Tenasserim Langur ■ *Trachypithecus barbei* HB 50–70cm, T 70–80cm

DESCRIPTION Dark greyish-black body and tail, slightly paler underside, grey face with large white rings around the eyes, and bare, whitish skin around the mouth. **DISTRIBUTION** Myanmar (east) and Thailand (west). Confined to a small area north of the Thai-Malay Peninsula. **HABITAT AND HABITS** A very poorly known species, and surveys are required to better understand its distribution, ecology and status. Current knowledge of it is derived from museum or zoo specimens, and very few observations of wild specimens. **Notes** Surveys are needed to better understand this species, its habitat requirements and conservation status.

Francois' Langur ■ *Trachypithecus francoisi* HB 47–63cm, T 40–97cm

DESCRIPTION Covered almost entirely in silky black hair, except for a white band from the mouth corners to above the ears, giving it an appearance of an extended moustache. Tall, pointed crest on the head. Infants are orange. **DISTRIBUTION** China (south-central) and Vietnam (north). **HABITAT AND HABITS** Lives in forests near and on limestone karsts, also taking shelter in caves and overhangs. Mainly eats leaves, flowers, shoots, bark and fruits. Calls loudly to establish territory. **Notes** This species is highly threatened, especially by illegal hunting.

Cat Ba Langur ■ *Trachypithecus poliocephalus* HB 49–59cm, T 80–90cm

DESCRIPTION Dark chocolate-brown body with lighter coloured head, neck and rump, which ranges from pale yellow to bright gold. Pale grey band on the rump and upper thighs, forming a 'V' shape at the base of the long, black tail. Easily identified because no other langur with a pale head occurs in the same range. Infants are bright orange. **DISTRIBUTION** Vietnam (north, on Cat Ba Island and in Ha Long Bay). **HABITAT AND HABITS** Lives in forests around limestone karsts. Mainly eats leaves, shoots, fruits, flowers and bark. **Notes** The White-headed Langur (see opposite) is considered by some authorities as a subspecies of the Cat Ba Langur.

White-headed Langur ■ *Trachypithecus leucocephalus* HB 47–62cm, T 77–89cm

DESCRIPTION Body is dark brown. Head, neck, shoulders and upper chest are whitish to cream, with varied white markings on the hands, feet, lower legs and forearms. Tail is whitish, with a brown base extending partially up the length of the tail. Whitish crest with some hairs tipped with brown. **DISTRIBUTION** China (south – restricted to south-western Guangxi Province). **HABITAT AND HABITS** Found in and around limestone karsts and cliff faces, and in associated forests. Sleeps in caves and limestone cracks, presumably in order to protect itself against predators and the elements. Feeds largely on leaves, but also consumes other plant material such as flowers and fruits. **Notes** The greatest threat to this species is poaching for use as food and in traditional medicines. Habitat loss is also a major threat, with forests being cleared for agriculture, and trees being cut for firewood and other purposes.

White-headed Langur

White-headed Langurs

Delacour's Langur ■ *Trachypithecus delacouri* HB 57–73cm, T 73–97cm

DESCRIPTION Glossy black head and body, with a highly contrasting white rump and outer thighs – locals refer to it as the 'langur with white trousers'. Black face with long grey hairs on the cheeks, and has a crest. Very long, black, thickly furred tail. **DISTRIBUTION** Vietnam (north-central). **HABITAT AND HABITS** Lives in forests on limestone karsts, taking shelter in caves. Mainly eats leaves, shoots, fruits and flowers. **Notes** Less than 250 individuals survive, with each of the 19 sub-populations being no greater than 50. More than half of the sub-populations do not exceed 20 individuals each. The decline of this Critically Endangered primate continues, because it suffers from poaching and habitat loss. The largest sub-population occurs in the Van Long Nature Reserve, which is a protected area guarded by enforcement rangers.

Lao Langur ■ *Trachypithecus laotum* HB 46–53cm, T 80–90cm

DESCRIPTION Glossy black body and head, with long white whiskers on the sides of the mouth, stretching to above the ears and around the back of the head, and nearly reaching the nape. Broad white band on the forehead and black crest. **DISTRIBUTION** Laos (west-central). **HABITAT AND HABITS** Lives only in limestone karst forests, in the Khammouane and Borikhamxai Provinces, though there are unconfirmed reports of it occurring north of the Nam Theun River. **Notes** Very little is known about this species. Surveys and research are required to learn more about its status and ecology.

Hatinh Langur ■ *Trachypithecus hatinhensis* HB 50–66cm, T 81–87cm

DESCRIPTION Glossy black fur overall, except for a white moustache extending from the sides of the mouth over the ears to the nape. Distinct black crest. Juveniles have a white band on their foreheads. Fairly similar to Francois' Langur and the Lao Langur (see pp. 78 and opposite), but the distinguishing characteristics are that the former's moustache does not continue behind its ears and the latter has a white forehead even in adults. **DISTRIBUTION** Laos (east-central) and Vietnam (north-central). **HABITAT AND HABITS** Spends considerable time in trees and on the ground. Mainly eats leaves. Lives in forests near limestone karsts and outcrops in rocky mountainous areas, though its previous range may have included a wider habitat variety.

Black Langur ■ *Trachypithecus ebenus* HB 62cm, T 94cm

DESCRIPTION Completely black with a slight brownish tinge. Has a crest on top of the head. **DISTRIBUTION** Laos (east-central) and Vietnam (central). **HABITAT AND HABITS** Lives on steep and high limestone cliffs, where it sleeps at night in caves, presumably for protection against predators and the weather. Feeds largely on leaves, but might take other plant material. **Notes** Very little is known about this species, and little or no research has been carried out on it in the wild. A great deal of work is needed to better understand and protect it.

Northern Plains Grey Langur ■ *Semnopithecus entellus* HB 41–78cm, T 80–118cm

DESCRIPTION Greyish-brown back and outer sides of limbs, yellowish-brown sides and an orangish-brown underside. Head is light greyish to creamy-yellow, and there are long, whitish whiskers. Tail is yellowish-brown with a white tip, and the hands and feet are black. Bare skin of the feet, hands, face and ears is black. **DISTRIBUTION** India (west to east across the country, north to the foothills of the Himalayas, and south to the Krishna River in Andhra Pradesh State). **HABITAT AND HABITS** Found in agricultural areas, secondary scrub forests, dry deciduous and thorny forests, and dry woodland. The vast majority of the population of this species lives in heavily altered, human-dominated areas, with few of the langurs actually found in forests. Frequently found close to human settlements. Food includes leaves, flowers, buds, seeds and a wide variety of other plant material, as well as insects and insect larvae. Also raids cultivated crops for food. **Notes** Numerous in parts of its range, but threatened by habitat loss and perhaps some hunting. Also killed by humans protecting their crops.

Kashmir Grey Langur ■ *Semnopithecus ajax* HB 51–79cm, T 69–102cm

DESCRIPTION Large-bodied langur, with yellowish-white fur and a black face and ears. Also has dark forearms and a long tail. Males are substantially larger than females. **DISTRIBUTION** India (north-west, in Himachal Pradesh). **HABITAT AND HABITS** Found in pine, moist temperate and alpine cedar forests at 2,200–4,000m, eating mainly leaves. Semi-terrestrial, moving quadripedally both on the ground and when in the trees. Occupies the small Chamba Valley of the state of Himachal Pradesh surrounded by high peaks, which is threatened by agriculture and other human settlement factors. **Notes** The remaining population is small, estimated to be less than 250 mature individuals. May also occur in the Kishtwar Valley of Jammu and Kashmir state.

Terai Grey Langur ■ *Semnopithecus hector*

HB 58–76 cm, T 75–99cm

DESCRIPTION Greyish-yellow to light sandy-brown body, long, thick hair covering the upper body, and dark dorsal line. Underside is whitish, as are the hands and feet, and the head is white. Tip of the tail is white and the bare skin on the hands, feet and face is black. **DISTRIBUTION** Bhutan (not confirmed – most probably occurs in south-west), India (north, in Himalaya foothills) and Nepal (west and central). **HABITAT AND HABITS** Lives in semi-evergreen and deciduous forests, as well as cedar and oak foothill forests, at 150–1,600m, in the Terai region. Survives in seasonally harsh weather, during which temperatures are frequently below freezing. Feeds on leaves, flowers, seeds, fruits, bark and other plant material. Like other species of this genus, it is mostly arboreal but does spend considerable amounts of time on the ground. **Notes** This species is threatened for a variety of reasons, all relating to habitat loss, including mining, logging, firewood collection and expansion of human settlements.

Central Himalayan Sacred Langur ■ *Semnopithecus schistaceus*

HB 58–76cm, T 75–99cm

DESCRIPTION Large langur, reaching as much as 20kg in weight. Long brown fur on the back and outer limbs, with the underside of the limbs and rump being whitish. Brown tail is tipped with white. Bare skin on the hands, feet and face is black. **DISTRIBUTION** Bhutan (west), China (south), India (north), Nepal and Pakistan (north-west). **HABITAT AND HABITS** Inhabits broadleaved and semi-evergreen sal forests, alpine and montane forests, and scrub forests in hilly areas, at 2,000–4,000m, in areas with seasonally harsh weather conditions. Spends a considerable amount of time in trees, but also on the ground, including on rocky outcrops. Feeds largely on leaves, but also consumes buds, seeds, flowers, fruits, nuts, bark, lichens and other plant material. **Notes** Habitat loss is the major threat to this species, with trees being harvested for firewood, charcoal production and other purposes. Hunting for food and for its parts, which are used in traditional medicines, is also a threat in some parts of its range. The species may occur in east Afghanistan, but this is yet to be confirmed.

Black-footed Langur ■ *Semnopithecus hypoleucos* HB 50–69cm, T 72–109cm

DESCRIPTION Purplish-brown with a pale-orange underside. Black limbs, face and ears, and long, bristly eyebrows. Long black tail with several white hairs on the tip. A medium-sized langur, though its size varies between the recognized subspecies. **DISTRIBUTION** India (west and south-west). **HABITAT AND HABITS** Lives in tropical, moist deciduous and riparian forests, and gardens and sacred groves, in the Kerala coastal region of south-west India. Specialized inhabitant of wet-evergreen and semi-evergreen montane forests. When leaves are scarce, eats flowers, buds, fruits and bark.

Tufted Grey Langur ■ *Semnopithecus priam* HB 41–78cm, T 75–86cm

DESCRIPTION Large-bodied, with silky light grey fur and a black face and ears. Long, dark grey tail and lanky limbs, with a pronounced fringe. Males (weighing 18kg) are significantly larger than females (11kg). **DISTRIBUTION** India (south, south-central and south-east), and Sri Lanka (dry zone, from Jaffna in the north, to the coast in the south). **HABITAT AND HABITS** In Sri Lanka, found in the scrub forests of the dry zone, including some urban areas. In India, occurs in the southern Western Ghats. Semi-terrestrial, it mainly eats leaves, flowers and fruits. Lives in large family groups of up to 60 individuals. **Notes** Often referred to as the Hanuman Langur, named after the Monkey God. It is said that the dark hands and feet are the result of Hanuman getting burnt while rescuing Sita, Lord Rama's wife, from the evil clutches of Ravana. With the strong links to Hindu beliefs, it is considered sacred. It can be easily seen in Yala National Park.

Purple-faced Langur ■ *Semnopithecus vetulus* HB 49–65cm, T 69–75cm

DESCRIPTION Heavily built, with dark, silky hair on the back, pale cheek whiskers and paler brown hair on the crown. Face is purplish-black, giving it its name. Juveniles are silvery-grey with a brownish tinge. **DISTRIBUTION** Sri Lanka (west, south-west, central, north and east). **HABITAT AND HABITS** Found in wide variety of habitats, from lowland forests to the highlands. Main diet comprises leaves, fruits, flowers and seeds. Lives in harem groups, with one male in groups of 6–20 individuals. Males are very territorial. **Notes** There are four subspecies, each occurring in different parts of Sri Lanka and having varying colouration. They are still easily identifiable as each occurs in different regions, and cannot be mistaken for Tufted Grey Langurs (see p. 93), Sri Lanka's only other langur. The Western Purple-faced Langur *S. v. nestor* ranks among the world's most threatened primates.

Nilgiri Langur ■ *Semnopithecus johnii* HB 51–65cm, T 76–97cm

DESCRIPTION Black to blackish-brown body, with a light brown head and nape, and a dark face. Rump and base of the tail are sometimes grizzled, and females have a white patch on the inner thighs. Young are reddish-brown until they turn black at approximately 10 weeks of age. **DISTRIBUTION** India (south-west, confined to Karnataka, Kerala and Tamil Nadu). **HABITAT AND HABITS** Found in evergreen, deciduous montane and riverine forests, ranging from 300m to more than 2,000m. Sometimes enters teak plantations. Feeds largely on leaves, but will also take seeds, flowers, buds and occasionally fruits. It has also been known to raid crops. Found in small groups of up to 25 individuals, it uses a wide range of vocalizations, from barks to whistles and whoops for communication. **Notes** This species is hunted for its body parts, which are used in local traditional medicines, for its meat, and to a lesser extent, live for the pet trade – it is one of the most heavily hunted primates in South Asia. Habitat loss and fragmentation is a major threat as well. Much research is needed to better understand this species' conservation requirements.

Tonkin Snub-nosed Monkey ■ *Rhinopithecus avunculus* HB 51–62cm, T 56–92cm

DESCRIPTION Brown to black back and tail, with yellowish-white to orange underparts, elbows, crown, ears and tail-tip. Blue-grey facial skin with pink lips. Infants are greyish-white. As in the case of all snub-nosed monkeys, its most prominent feature is its tiny nose with forwards-facing nostrils. **DISTRIBUTION** Vietnam (north-west), and possibly neighbouring China. **HABITAT AND HABITS** Lives in primary forests in limestone karsts, feeding mainly on young leaves, unripe fruits and seeds. **Notes** This Critically Endangered species has an extremely restricted range, numbering approximately 250 individuals, due to extreme poaching pressure and habitat loss.

Black Snub-nosed Monkey ■ *Rhinopithecus bieti* HB 74–83cm, T 51–72cm

DESCRIPTION Greyish-black chest, limbs and upper back, with whitish lower back, hips and thighs, and yellowish-white cheeks, throat, ears and abdomen. Pink facial skin, dark brow and tufted tail. **DISTRIBUTION** China (south-west – Yunnan and Tibet). **HABITAT AND HABITS** Found in high-altitude evergreen forests, and with its long, dense hair it is well adapted to cold climates. Mainly feeds on lichen that grows on evergreen trees, bark, grasses, leaves and berries. Lives in large bands of up to 200 individuals, comprising smaller groups.

Black Snub-nosed Monkey

Grey Snub-nosed Monkey

Rhinopithecus brelichi HB 64–69cm, T 70–85cm

DESCRIPTION Greyish-brown back, whitish-grey abdomen, pale grey thighs, and yellowish-orange upper arms and brow. Brown crown, white ear-tips, blue facial skin and pink lower lip. Males have a small white patch between the shoulders. **DISTRIBUTION** China (south-central). **HABITAT AND HABITS** Found in the Guizhou province in southern China, restricted to the mountain forests on and around Fanjing Mountain in the Wuling range. Mainly eats leaves, buds, fruits, bark and lichen, as well as insect larvae. **Notes** With under 1,000 remaining individuals found in a single locality, this species is vulnerable to factors that continue to threaten its survival, including forest loss for development, agriculture and firewood collection, as well as non-targeted hunting.

Golden Snub-nosed Monkey ■ *Rhinopithecus roxellana* HB 52–78cm, T 57–80cm

DESCRIPTION Stunning shaggy coat ranging from yellowish-orange to reddish-golden on the limbs and body. Some dark brown or black on the crown and back, extending to the tail. White, hairless muzzle with pale blue skin around the eyes. Infants are pale grey, sometimes almost white. Adult males have a wart-like growth at the mouth corners, the sole example of such a feature in primates. The purpose of these growths has not been determined. **DISTRIBUTION** China (west-central). **HABITAT AND HABITS** Found only in temperate montane forests at 2,000–3,500m, descending to mixed and broadleaved forests in winter. Clear preference for undisturbed primary forest. Mainly feeds on lichen, leaves, hemlock shoots and bark.

Myanmar Snub-nosed Monkey ■ *Rhinopithecus strykeri* HB 56cm, T 78cm

DESCRIPTION Almost completely black, with white ear-tufts, bearded chin and upturned nose, as well as some white hairs above the upper lip, giving it the appearance of a moustache. It has a thin and high forwards-curved black crest, with pink facial skin. Inner sides of the upper arms and legs are blackish-brown. Very long tail, approximately 140 per cent of the body length. **DISTRIBUTION** Myanmar (north-east) and China (south-west, in the Gaolilgongshan Nature Reserve). **HABITAT AND HABITS** Found in cool temperate rainforests and mixed temperate forests in steep, mountainous areas. Lives in the north-east Kachin state in Myanmar, isolated from other species by the Mekong and Salween Rivers. **Notes** Only discovered in 2010, this species is Critically Endangered, but there are no specific conservation measures in place for it. In Myanmar, the area where it occurs is not yet under government control, and therefore there is an absence of border control and other enforcement agencies.

Grey-shanked Douc ■ *Pygathrix cinerea* HB 61–76cm, T 56–76cm

DESCRIPTION Pale grey body, abdomen, crown and arms; medium to dark grey legs; and black shoulders, upper legs and part of the rump. Hairless yellowish-brown facial skin with white whiskers on sides of the face. White tail with a thin tassel, and a white patch on the rump at the base of the tail. Almond-shaped eyes. **DISTRIBUTION** Vietnam (central), Cambodia (north-east) and probably southern Laos. **HABITAT AND HABITS** Lives in both primary and disturbed evergreen and semi-evergreen forests. Mainly eats young, tender leaves, buds, fruits, seeds and flowers. Feeds high in treetops. Highly social, it lives in groups that vary in size according to habitat quality and food availability. **Notes** This Critically Endangered primate, the rarest of the doucs, occurs only in the central highlands of Vietnam. It is estimated that only 550–700 individuals survive, although some sites with assumed occurrence have yet to be surveyed.

Red-shanked Douc ■ *Pygathrix nemaeus* HB 61–76cm, T 56–76cm

DESCRIPTION A strikingly colourful, large primate. Dark reddish-chestnut lower legs, and black hands, feet, shoulders, insides of upper arms, upper legs and rump; white lower arms and speckled grey back, belly and tops of the upper arms. Yellow-brown face and a long white tail. **DISTRIBUTION** Cambodia (small area in the north-east), Laos (east-central and south-east), Vietnam (north and central – very fragmented). **HABITAT AND HABITS** Found in tall evergreen and semi-evergreen primary forests, and lowlands up to 2,000m, including limestone outcrops. Mainly arboreal, and feeds on leaves and buds, as well as some fruits, flowers and seeds. When relaxed, it moves noisily in the trees, disappearing quietly only when disturbed.

Black-shanked Douc ■ *Pygathrix nigripes* HB 61–76cm, T 56–76cm

DESCRIPTION Dark-speckled grey crown and upperparts, paler grey underparts, white chin, throat and tail, and black limbs with paler frosting on the arms. Blue-grey facial skin with distinct yellow-orange eye-rings. **DISTRIBUTION** Cambodia (east) and Vietnam (south-west), and possibly Laos. **HABITAT AND HABITS** Lives in evergreen, semi-evergreen and mixed deciduous forests. Feeds on leaves, seeds, fruits, flowers and buds.

Proboscis Monkey

Nasalis larvatus HB 55–65, T 62–74cm

DESCRIPTION Best known for its oversized nose, the Proboscis Monkey is a very large primate, with males reaching up to 20kg in weight. Adult males have large stomachs. Reddish-brown fur and greyish limbs. Females are significantly smaller than males, and like the juveniles, they have small, upturned noses. **DISTRIBUTION** Endemic to Borneo. Brunei, Indonesia (Kalimantan) and Malaysia (Sabah and Sarawak). **HABITAT AND HABITS** Inhabits riparian-riverine forests and coastal lowland forests, including mangrove, peat-swamp and freshwater swamp forests. Mainly eats young leaves and unripe fruits. It has partly webbed back feet, which aid it in balancing on mangrove mud and swimming.

Proboscis Monkey

Pig-tailed Langur ■ *Simias concolor* HB 46–55cm, T 14–15cm

DESCRIPTION This stocky primate occurs in two colour morphs; dark grey, which is most common, and creamy-buff. It has a black face with a small snub nose. The short, curly tail is hairless except for a small tuft at the end. **DISTRIBUTION** Indonesia (Mentawai Islands off the western coast of Sumatra). **HABITAT AND HABITS** Lives in lowland tropical forests, including hillsides in primary forests. Mainly eats leaves, unripe fruits and flowers. **Notes** Locally known as *simakobu*, this is one of the world's most threatened primates. It is imperiled by heavy hunting and serious loss of habitat due to commercial logging and forest conversion for oil-palm production.

CERCOPITHECIDAE (MACAQUES)

This highly successful group of primates is the most widespread with the exception of humans. Macaques are found on three continents globally: Asia, Africa and Europe (historically present in the latter case, though now only an introduced population remains). They live in a variety of habitats in uplands and lowlands. Of the 22 species currently recognized globally, Asia is home to 21. The only macaque that occurs outside Asia is the Barbary Macaque *Macaca sylvanus*.

As far as primate watching goes, macaques are among the most rewarding and entertaining, because they are gregarious, generally living in large troops and foraging on the ground as well as in trees. Many species are also easy to find and watch due to a degree of habituation to human presence, and they are often found living near temples and tourist spots. Some people give them food, but this is strongly discouraged because it leads to the macaques becoming more aggressive, stealing food and often hurting people in the process. These macaques then earn the label of 'problem animals', and the authorities have to solve a problem that has no easy solution and often results in unnecessary killing.

Macaques are diurnal, omnivorous and can live to 30 years of age. They have interesting and complex social hierarchies, and communicate extensively using facial expressions, calls and touch.

Lion-tailed Macaque ■ *Macaca silenus* HB 46–61cm, T 25–38cm

DESCRIPTION Black body, limbs, tail, hands, feet and face. The most distinctive features, however, are the long, lion-like tail, which ends with a tuft, and a long mane of grey to brownish-grey hair growing from the sides of the head and cheeks. **DISTRIBUTION** India (in the southern Indian states of Karnataka, Kerala and Tamil Nadu, in the Western Ghats). **HABITAT AND HABITS** Found in dense evergreen forests, sometimes entering disturbed forests and fruit plantations, in hilly areas up to 1,600m. Prefers the forest canopy, feeding on fruits, seeds, flowers and other plant material, and on insects and other invertebrates. **Notes** Highly threatened by forest fragmentation, due largely to logging and the expansion of commercial crops. It is also hunted and eaten by locals in some areas.

Northern Pig-tailed Macaque ■ *Macaca leonina* HB 47–59cm, T 14–23cm

DESCRIPTION Similar to the Southern Pig-tailed Macaque (see p. 112), but with a lighter build and shorter muzzle. Brown to golden-brown fur with a lighter, partially bare face. Distinctive red streaks on the face pointing diagonally upwards from the eyes. **DISTRIBUTION** Bangladesh (eastern), Cambodia, China (south-western Yunnan), India (the states of Arunachal Pradesh, Assam, Manipur, Meghalaya, Mizoram, Nagaland and Tripura), Laos, Myanmar including the Mergui islands, Thailand (throughout the country but only extending about halfway down the peninsula) and Vietnam (south). **HABITAT AND HABITS** Lives in a variety of forest habitats, including disturbed forests. Feeds mainly on the ground, taking to the trees to escape when threatened. **Notes** Closely related to the Southern Pig-tailed Macaque, and with the slight overlap in range some natural hybridization occurs.

Southern Pig-tailed Macaque ■ *Macaca nemestrina* HB 47–59cm, T 14–23cm

DESCRIPTION Stocky, heavy-set macaque with a short, curly tail and an olive-brown coat, a dark brown crown and whitish underparts. **DISTRIBUTION** Brunei, Indonesia (Kalimantan and Sumatra), Malaysia and Thailand (southern peninsula). **HABITAT AND HABITS** Often found in hilly areas, foraging largely on the ground. Diurnal and widespread. Eats fruits and small animals. Usually lives in large groups of 15–40 individuals, but males are sometimes solitary.

Siberut Macaque ■ *Macaca siberu* HB 47–48cm, T 7–8cm

DESCRIPTION Stocky in appearance, with a short tail. Mostly black above and with a light greyish-brown underside and whitish throat. Arms and flanks are brown. Cheeks are pale silver but the rest of the face, hands and feet are black. Infants are whitish with reddish skin on the face, hands and feet, but are already more adult-like in colouration within a few weeks. **DISTRIBUTION** Indonesia (endemic to the island of Siberut, off the west coast of Sumatra). **HABITAT AND HABITS** Found mostly in dense continuous forests, including mixed dipterocarp forests, riverine and nipa palm forests, and mangroves. Both primary and secondary forests are used, from sea level up to 384m (the highest elevation on the island). Largely frugivorous, but leaves, mushrooms and flowers also form part of its diet, as well as occasionally invertebrates such as ants, termites and spiders. Very occasionally, consumes crabs and shrimp from rivers. **Notes** The Siberut Macaque was once considered a subspecies of the Southern Pig-tailed Macaque (see opposite), and much of its behaviour is similar. It is threatened by hunting, and found in lower densities near human settlements. Habitat loss due to commercial logging and conversion of forests to oil-palm plantations is also a serious threat.

Pagai Island Macaque ■ *Macaca pagensis* HB 40–55cm, T 10–16cm

DESCRIPTION Back and legs are dark brown, while the arms are reddish-brown. Underside, throat and sides of the neck across to the front of the shoulders are pale brown to chestnut. Bare facial skin is black. **DISTRIBUTION** Indonesia (the southern Mentawai Islands off the west coast of Sumatra: Pagai Selatan, Pagai Utara and Sipora). **HABITAT AND HABITS** Found in a variety of habitats, including coastal swamp and riverine forests. Occurs in primary and secondary forests. Feeds on fruits, especially figs, and other plant material, often in the company of Mentawai langurs. Normally sleeps high in the forest canopy. **Notes** Until recently, this species was included in Siberut Macaque (see p. 113), which was regarded as a subspecies of the Southern Pig-tailed Macaque (see p. 112). It is seriously threatened by hunting and capture for the pet trade, and by habitat loss due to logging and conversion of forests to oil-palm plantations.

Black-crested Macaque ■ *Macaca nigra* HB 45–57cm, T 25cm

DESCRIPTION Fur is entirely black, as is the bare skin on the face and hands. Very short tail, and the bare pads on the rump are pink. Cheek bones and brow are very prominent and bony. Hair on the head is long and forms a crest. **DISTRIBUTION** Indonesia (north-east Sulawesi and the adjacent islands of Manado Tua and Talise). **HABITAT AND HABITS** Found in both primary and secondary rainforests, montane forests, mangroves and heavily cultivated areas, up to 1,350m. Feeds on a wide variety of plant matter, including leaves, fruits, seeds and flowers, as well as fungi and some animal matter, such as small vertebrates, insects and birds' eggs. Also raids crops. The majority of its time is spent on the ground, foraging, grooming and travelling. **Notes** Hunting for consumption, and to a lesser extent for the pet trade, is a severe threat to this species. The meat of the species is traded in local markets, despite the animal being totally protected in Indonesia. Habitat loss is also a serious threat. Tangkoko Nature Reserve in northern Sulawesi is an excellent place in which to see it.

Black-crested Macaque

Gorontalo Macaque ■ *Macaca nigrescens* HB 50–60cm, T 2.5cm

DESCRIPTION Overall dark reddish-brown to black, with a black streak on the lower back. Distinct crest of elongated hairs on the head, making it look very similar to the Black-crested Macaque (see p. 115), although its crest is slightly shorter and the upper half of its body is lighter. Juveniles are pale brown with blackish hands, feet and forearms.
DISTRIBUTION Indonesia (central section of the northern peninsula of Sulawesi).
HABITAT AND HABITS Lives in rainforests at moderate elevations, and in hill forest up to 2,000m. Feeds largely on fruits, but also eats leaves and other plant material, including vegetables that it raids from cultivated areas, and some invertebrates. **Notes** Previously considered a subspecies of the Black-crested Macaque, it is believed to hybridize with it on the western side of Mount Padang and with Heck's Macaque (see p. 120) east of the Bolango River and near Bolaangitang. This species is hunted and eaten by locals, especially during the Christmas season.

Tonkean Macaque ■ *Macaca tonkeana* HB 50–68cm, T 3–7cm

DESCRIPTION Black to dark grey with areas of lighter brown to buff on the rump, cheeks and brow. Tail is inconspicuous. Bare skin on the rump is pinkish and on the face it is black. **DISTRIBUTION** Indonesia (central Sulawesi and Togian Islands). **HABITAT AND HABITS** Found in primary and secondary rainforests, from sea level to 2,000m. Feeds on plant material, including fruits, seeds and young leaves, as well as on some invertebrates. Also known to raid crops and eat a variety of cultivated fruits and vegetables. As its habitat diminishes, it is increasingly reliant on these cultivated food sources. Spends considerable time in the forest canopy – perhaps more so than other Sulawesi macaques. Where ranges overlap, it may hybridize with Booted, Black-crested and Heck's Macaques (see opposite and pp. 115 and 120). **Notes** Threatened by habitat loss due to conversion of forest to plantations, and persecuted for raiding crops. Also hunted and eaten, and captured as pets.

Booted Macaque ■ *Macaca ochreata* HB 50–59cm, T 35–40cm

DESCRIPTION Adults are black with grey hind legs and forearms. Hands and feet are often darker grey, and there is a lighter grey rump-patch. Facial skin is black, and it has a short tail. **DISTRIBUTION** Indonesia (endemic to south-east Sulawesi, throughout the entire south-east peninsula, and the adjacent islands of Muna and Butung). It may no longer occur on Muna. **HABITAT AND HABITS** Found in forests up to 800m. Largely eats fruits, but also young shoots and leaves, and occasionally raids cultivated fruit and vegetable orchards. **Notes** Severely threatened by habitat loss, largely for plantations and other cultivation, mining and expanding human settlement. Sometimes poisoned for crop raiding.

Heck's Macaque ■ *Macaca hecki* HB 50–68cm, T 2–3cm

DESCRIPTION Black back with paler brownish underside. Arms are dark brown and legs are paler brownish-grey. Short tail, and bare patches on the rump are greyish. Bare facial skin is black. **DISTRIBUTION** Indonesia (north-western Sulawesi). **HABITAT AND HABITS** Found in primary and secondary rainforests. Feeds largely on fruits, but also eats leaves, flowers and some invertebrates. Also raids crops for maize, fruits and vegetables. One of the least-studied macaques, with very little known of its ecology, populations and conservation status. Hybridizes with the Tonkean Macaque (see p. 118) and possibly with the Gorontalo Macaque (see p. 117) in areas where their ranges overlap, with the offspring sharing external characteristics of both parents. **Notes** The greatest known threat to the continual survival of this species is the fragmentation and loss of its habitat due to agriculture and plantation expansion. Also persecuted as an agriculture pest.

Moor Macaque ■ *Macaca maura* HB 50–69cm, T 2–3cm

DESCRIPTION Overall dark brown to black fur, and a pale brownish-grey rump-patch with pink bare skin on the rump. **DISTRIBUTION** Indonesia (south-western peninsula of Sulawesi). **HABITAT AND HABITS** Found in forest and grassland mosaics, rainforests and deciduous forests, and on karst outcrops, from sea level up to 2,000m. Feeds largely on fruits, but also eats seeds, leaves and other plant materials, as well as some invertebrates. In areas where it overlaps with the Tonkean Macaque (see p. 118), hybridization occurs naturally. **Notes** The major threat to this species is loss and fragmentation of habitat due to agricultural and human settlement expansion. The destruction of karst outcrops through mining for cement is also a threat to the species, which often seeks refuge in these areas, especially when other available habitat has been destroyed. It is additionally persecuted as an agricultural pest by farmers, as well as being captured by local people seeking to keep it as a pet.

Toque Macaque ■ *Macaca sinica* HB 43–53cm, T 46–62cm

DESCRIPTION The body, limbs and long tail are brown to golden-brown, and there is a cap of hair radiating from the centre of the head. This hair is brown with a red tinge in lowland races, and without the reddish tinge in the northern race. The facial skin is pale, though slightly more reddish in the female than in the male. **DISTRIBUTION** Sri Lanka (north-east and south-east, south-west, south-central). **HABITAT AND HABITS** Found in a wide range of forest types, from wet lowland forests to dry forests and tropical montane forests, from sea level to approximately 2,100m. Often occurs near water, and agricultural areas and human settlements. Varied and somewhat opportunistic diet that includes plant, insect and other animal matter, and occasionally food scavenged from refuse dumps. **Notes** Habitat loss, especially due to expansion of plantations and other agricultural enterprises, is the greatest threat to the Toque Macaque. It is also killed as a crop pest.

Bonnet Macaque ■ *Macaca radiata* HB 38–59cm, T 33–64cm

DESCRIPTION Greyish-brown back and a lighter underside, with a circular cap of hair on the head. Red bare skin on the face and a very long tail. **DISTRIBUTION** India (southern India). **HABITAT AND HABITS** Found in a wide variety of forest types, including scrub, evergreen and deciduous forests, plantations, cultivated land and even urban areas, up to 2,600m. Locally abundant and in some areas largely commensal, with greater populations in urban and cultivated areas than in forests. Feeds on plant material, including fruits, nuts, seeds and grains, as well as invertebrates. Raids crops and homes for food, and some populations feed heavily on foods given by humans. **Notes** Can be easily observed in cities, towns and farmland. Sometimes hunted and traded live for pets, entertainment and biomedical research.

Assamese Macaque ■ *Macaca assamensis* HB 51–74cm, T 15–30cm

DESCRIPTION Yellowish-grey to dark brown fur, with paler shoulders and limbs. Sparser hair on the underparts, making the bluish skin slightly visible. Dark hairs on the cheeks are directed backwards towards the ears, and the hair on the crown is parted in the middle. Dark brown to purplish facial skin, which is hairless. **DISTRIBUTION** Bangladesh (south-west), Bhutan, China (south-west), India (north-east), Laos, Myanmar (south and east, through north and east, Nepal (central), Pakistan, Thailand (north and west) and Vietnam (north). **HABITAT AND HABITS** Mainly lives in montane forests up to 3,500m, mostly eating fruits, leaves and small animals including insects.

Tibetan Macaque ■ *Macaca thibetana* HB 49–71cm, T 6–10cm

DESCRIPTION Long, thick dark brown fur on the back and lighter buff-white fur on the underside. Stocky in appearance with a short tail. Bare skin on the face is pink to red.

DISTRIBUTION China (east and central), and possibly north-east India. **HABITAT AND HABITS** Found in primary and secondary forests – mostly in broadleaved evergreen forests, but in subtropical forests as well. Lives in mountainous areas at 1,000–3,000m, where it sometimes sleep in caves. Foods eaten include fruits, berries, flowers, leaves and other plant material, as well as some invertebrates. Battles for dominance among males are frequent and can be violent, sometimes resulting in death. **Notes** In some areas this species is quite common, and in others it is increasingly common in tourist areas, where feeding the macaques should be discouraged.

Arunachal Macaque ■ *Macaca munzala* HB 58cm, T 26cm

DESCRIPTION Large, heavy-set macaque with a very dark brown lower back, hindlegs and tail, which is stocky. Dark brown face and crown-patch, with a pale yellow patch on the front of the crown. Upper part of the face is broader than the muzzle, and this is especially obvious in adult males. Juveniles have relatively hairless, whip-like tails that taper into a narrow tip. **DISTRIBUTION** India (north-east, possibly extending into Bhutan and west China). **HABITAT AND HABITS** Known from a variety of habitats, including primary oak and coniferous forests, secondary broadleaved forests, and degraded scrubland and agricultural areas, in high altitudes of western Arunachal Pradesh, at 1,800–3,500m. Feeds on fruits, leaves, shoots and seeds, and in the winter adopts a much more fibrous diet that includes mature leaves, bark and winter buds. **Notes** The population of this new species, described as recently as 2005, may be fewer than 500 individuals and declining. It is sometimes hunted for its meat and persecuted for crop raiding. Very little is known about it, and it may yet be found to occur in neighbouring Tibet (China) and Bhutan.

Arunachal Macaque

White-Cheeked Macaque ■ *Macaca leucogenys* HB 58–72cm, T 26–28cm

DESCRIPTION Light to dark brown with a lighter underside. Heavyset with a relatively short and sparsely haired tail. Dark facial skin on the muzzle, which is paler or reddish in infants. Adults and sub-adults have prominent white side-whiskers which extend from cheeks to ears, giving the species its name; and a thin dark stripe running from the outer corner of the eyes to the ear.

DISTRIBUTION China (south-eastern Tibet). **HABITAT AND HABITS** Observed in a variety of habitats, from tropical forests at 1,395m, to primary and secondary evergreen broadleaved forest and in mixed broadleaf-conifer forests, at 2,700m. **Notes** Very little is known about this species as it has only been recently described (2015), and more research is required. It is threatened by illegal hunting and habitat loss.

Stump-tailed Macaque ■ *Macaca arctoides* HB 49–64cm, T 4–8cm

DESCRIPTION Stout, heavy build, similar in general stocky appearance to that of the pig-tailed macaques, with shaggy reddish-brown to darker brown fur. Tail is very short and is usually held down (except when it is alarmed), making it appear tailless. The most striking feature is the bare facial skin, which ranges from dark pink to bright red; it becomes darker with age, and turnes bright red when the animal is excited. **DISTRIBUTION** Bangladesh (may be extirpated), Cambodia, China (south-west), India (north-east), Laos, Malaysia (extreme northern Peninsular Malaysia), Myanmar (north), Thailand and Vietnam. **HABITAT AND HABITS** Found in primary and secondary forests up to 2,000m. Main diet comprises seeds, fruits and buds, as well as small animals including insects. Lives in large troops, feeding mainly on the ground. **Notes** Last recorded in Bangladesh in the 1980s.

Long-tailed Macaque ■ *Macaca fascicularis* HB 45–55cm, T 44–55cm

DESCRIPTION Grey-brown to reddish-brown fur, with slightly paler undersides and a brownish-grey face. Lean build, with males being significantly larger than females, and a long tail. **DISTRIBUTION** Bangladesh (south-west), Brunei, Cambodia, India (Andaman and Nicobar Islands), Indonesia (Sumatra, Java, Bali and most but not all offshore islands in the Greater Sundas), Laos (southern), Malaysia, Myanmar (south), the Philippines, Singapore, Thailand (west-central, east and south, including offshore islands), Timor-Leste and Vietnam (south-east). **HABITAT AND HABITS** Especially common near coastal areas and forest edges, frequently near people. An omnivore, it eats a wide range of animal matter and vegetation. It is diurnal, sleeping in the branches of trees during the night. Gregarious species, often congregating in groups of 20–30 individuals, though sometimes groups number more than 50. **Notes** This is one of a few primate species that use tools: populations in southern Thailand and southern Myanmar use stones and shells to crack open marine molluscs and nuts. The only macaque in the Philippines.

Long-tailed Macaque

Rhesus Macaque ■ *Macaca mulatta* HB 47–59cm, T 21–28cm

DESCRIPTION Brownish upperparts with lighter brown underparts. Hair on the crown is short and directed backwards. **DISTRIBUTION** Bangladesh, Bhutan, China, India, Lao, Myanmar, Nepal, Pakistan, Thailand and Vietnam. Also found in Afghanistan. **HABITAT AND HABITS** Lives in large troops of up to 50 individuals, in secondary and disturbed forests, often in close proximity to human settlements. Naturally, its main diet includes fruits, seeds and small animals, but when close to humans it raids crops and is also fed by locals, especially near Hindu temples in India, where it is held sacred in some areas.

Japanese Macaque ■ *Macaca fuscata* HB 47–60cm, T 7–12cm

DESCRIPTION Uniformly greyish-brown with a bare red face and area around the anus. Tail is short. Fur thickens in winter, and in some areas this species lives in snowy conditions for a significant part of the year. **DISTRIBUTION** Japan (islands of Honshu, Shikoku and Kyushu, as well as Awaji, Shodo, Yaku, Kinkazan in Miyagi Prefecture and Kojima in Miyazaki Prefecture. **HABITAT AND HABITS** Widespread in many parts of Japan, and found in a variety of forest habitats, including broadleaved and evergreen forests, up to 1,500m. It occurs in subtropical conditions in the south to near subarctic conditions in the north, where it has adapted to live in snowy and freezing conditions. In some parts of its range, this species spends time in hot-spring pools for warmth. Feeds on fruits, including berries, nuts, leaves, seeds, bark, insects and small vertebrates. In some areas there are feeding associations between Japanese Macaques and Sika Deer *Cervus nippon*. **Notes** Raids agricultural lands in some places, and as a result large numbers of these macaques are culled each year to prevent damage to crops.

Formosan Rock Macaque ■ *Macaca cyclopis* HB 40–55cm, T 26–45cm

DESCRIPTION Uniformly brownish-grey with a lighter underside. Tail is of medium length and the facial skin is pinkish. **DISTRIBUTION** Taiwan, Province of China (mainly in central mountainous region, with some smaller populations in foothills and nearby lowlands). **HABITAT AND HABITS** Found in a wide variety of habitats, including broadleaved evergreen and conifer, bamboo and secondary forests, from sea level up to 3,600m. Feeds on fruits, including berries, seeds and other plant matter, as well as on insects and other animals. Sometimes enters agricultural areas seeking food, and is considered a pest by local farmers. In Taiwan, this species is sometimes preyed upon by the Mountain Hawk-eagle *Nisaetus nipalensis*. **Notes** Introduced and feral in parts of Japan, including Oshima, Nojima, Wakayama Prefecture and the Shimokita Peninsula.

HYLOBATIDAE (Gibbons)

These are the agile acrobats of the primate world, brachiating in a seemingly effortless fashion from branch to branch. They are also able to move bipedally, both on the ground and in the trees.

Most gibbons live in tropical forests, although the hoolocks are found in high montane forests. Gibbons eat primarily fruits, leaves, flowers and insects, and occasionally small vertebrates. They are often seen hanging from one branch with one hand while the other gathers food, which is eaten in that position. Gibbons are also very important seed dispersers.

Gibbon-inhabited forests echo with their loud, melodious, whooping, wailing songs. Most gibbons sing in male-female duets to mark territory; each species has a unique song, making it a useful tool in species identification in the field when direct sight observation is difficult.

Western Hoolock ■ *Hoolock hoolock* HB 45–65cm, T absent

DESCRIPTION Generally shaggy haired. Adult males and juveniles are mostly black, with thick, joined white eyebrows, and a white tuft on the chin or under the eyes. The preputial tuft is black or faintly grizzled. Females are coppery-buff-brown, with a white facial ring and dark brown cheeks. **DISTRIBUTION** Bangladesh, India (north-east) and Myanmar (north-west). **HABITAT AND HABITS** Found in primary and semi-evergreen forests. Fruits, leaves and shoots are the main diet. Females lead the movement of the group, which sometimes travels on the ground to reach fruiting trees, especially in degraded habitat. **Notes** Formerly conspecific with the Eastern Hoolock (see p. 139).

Gaoligong Hoolock ■ *Hoolock tianxing* HB presumably similar to Hoolocks, T absent

DESCRIPTION Males are dark brown, with thin white, clearly separated eyebrows (thinner than the Eastern Hoolock's, see p. 139). Black or brown beard and dark grey or black preputial tufts. Females are yellowish-white to reddish-blonde with sparse and incomplete face and eye-rings. Juveniles lack the white hair on the chin and under the eyes; eyebrows are not always clearly separated. **DISTRIBUTION** China (south-western) and Myanmar (eastern), between the Irrawaddy-Nmai Hka and the Salween Rivers. **HABITAT AND HABITS** Has been recorded from humid montane evergreen broad-leaved forests. More research is required to understand the behaviour of this species, but it is presumably similar to other hoolocks. **Notes** More research is essential to assess the status of this newly described species (description published in 2016), but it is thought to be low and, like other hoolocks, threatened by habitat loss, hunting and illegal trade.

Eastern Hoolock ■ *Hoolock leuconedys* HB 45–65cm, T absent

DESCRIPTION Similar in appearance to the Western Hoolock (see p. 136), but mainly distinguished by its thick white eyebrows, which are not joined, and its white preputial tuft. **DISTRIBUTION** China (Yunnan), India (north-east) and Myanmar (eastern). **HABITAT AND HABITS** Found in closed canopy, lowland broadleaved semi and mixed evergreen forests; occurs up to 2,700m in Yunnan. Though specific information on its feeding habits is unknown, they are most probably similar to those of the Western Hoolock. **Notes** Myanmar holds the largest population of this species, with the majority of these hoolocks being found in Kachin Northern Forest Complex and Southern Kachin – Northern Sagaing forests. Formerly considered a subspecies of the Western Hoolock.

Pileated Gibbon ■ *Hylobates pileatus* HB 47–60cm, T absent

DESCRIPTION Males are covered in short black fur, and have white brows, hands and feet, and a circular white streak around the cap on the sides of the head. Females are shaggier, from buff to silvery-grey, with a black cap, chest and cheeks. Subadult males are also silvery-buff, but become black at maturity, making this the only *Hylobates* species where males have a change in colour. **DISTRIBUTION** Cambodia (western – west of the Mekong River), Laos (south-western) and Thailand (south-eastern). **HABITAT AND HABITS** Inhabits mixed deciduous evergreen forest. Mainly eats fruits, young leaves and shoots, and insects. Calls loudly, with the female starting the song and the male joining in halfway through the female's final trilling. **Notes** In Thailand's Khao Yai National Park, it is sympatric and is known to hybridize with the White-handed Gibbon (see p. 142).

White-handed Gibbon ■ *Hylobates lar* HB 45–60cm, T absent

DESCRIPTION Blond or dark-brown forms can occur in the same family, unrelated to sex. It has long limbs, with white feet and hands. There is a pale ring encircling its face.

DISTRIBUTION China (south – possibly extinct), Indonesia (north Sumatra), northern Laos (north-west of Mekong), Malaysia (Peninsular Malaysia), Myanmar (south and east) and Thailand (south-west and north-west). **HABITAT AND HABITS** Often seen dangling from branches or huddled in forks of trees. Loud and distinct, high-pitched, whooping call. Mainly eats fruits, young shoots, leaves and insects. An important seed disperser as it swallows nearly all the seeds in its food.

Agile Gibbon ■ *Hylobates agilis* HB 45–65cm, T absent

DESCRIPTION Occurs in two colour forms, buff-blond and very dark brown. In mainland South-east Asia the dark brown phase is more common. It has a pale brow and cheeks, with a dark face. **DISTRIBUTION** Indonesia (Sumatra – south-east of Lake Toba and the Singkil River), Malaysia (Peninsular Malaysia – from the Mudah and Thepha Rivers in the north to the Perak and Kelanton Rivers in the south) and Thailand (southernmost Thailand, near the Malaysian border). **HABITAT AND HABITS** Lives in small family groups in tall dipterocarp forests, rarely descending to the ground. **Notes** Severely threatened by habitat loss due to logging and conversion to oil palm, and by the illegal wildlife trade.

Bornean White-bearded Gibbon ■ *Hylobates albibarbis* HB (no specific measurements available), T absent

DESCRIPTION Usually has dark brown upperparts, chest and head-cap, with black hands and feet, and brown fore and hind limbs. Not all individuals have white eyebrows and/or cheeks, but this characteristic is common. Infants are pale brown. **DISTRIBUTION** Indonesia (Kalimantan, south of the Kapuas River and west of the Barito River). **HABITAT AND HABITS** Lives in primary and disturbed secondary forests, as well as lowland and montane areas, including peat swamps. Main diet consists of fruits, and also insects and young leaves. **Notes** Before 2001, this was considered a subspecies of the Agile Gibbon (see p. 143). It is found in several protected areas, including the Tanjung Puting National Park, which is well known for its orangutans. The largest remaining population is in Sabangau, central Kalimantan. The species is known to hybridize with other gibbon species within its range.

Mueller's Bornean Gibbon ■ *Hylobates muelleri* HB 42–47cm, T absent

DESCRIPTION Fur colour varies from grey to brown. Dark-coloured cap and dark chest and underparts. Pale cream brows, which sometimes extend to appear like an incomplete facial ring. **DISTRIBUTION** Indonesia (south-east Kalimantan). **HABITAT AND HABITS** Found in primary and secondary forests, and in selectively logged forest. Favours high-sugar, fleshy fruits, but also eats young leaves and small insects.

Abbott's Grey Gibbon ■ *Hylobates abbotti* (No specific body measurements available)

DESCRIPTION Overall medium grey body, hands and feet, with a darker crown and ventrum. Until recently considered as a subspecies of Mueller's Bornean Gibbon (see p. 145), and now regarded as a full species based on genetic and morphological differences. **DISTRIBUTION** Endemic to Borneo. Indonesia (west Kalimantan) and Malaysia (Sarawak). **HABITAT AND HABITS** Found in primary and secondary semi-deciduous monsoon, dipterocarp and tropical forests in south-west Borneo. The main diet consists of fruits, leaves, flowers and insects. Hybridizes with the Bornean White-bearded Gibbon (see p. 144) in central Borneo.

East Bornean Grey Gibbon ■ *Hylobates funereus* HB 48–49cm, T absent

DESCRIPTION Very dark brown or grey body, with hands and feet being the same colour or lighter. Hair on the crown is elongated over the ears. Until recently it was considered as a subspecies of Mueller's Bornean Gibbon (see p. 145). **DISTRIBUTION** Endemic to north and north-east Borneo. Indonesia (east Kalimantan) and Malaysia (Sabah and Sarawak). **HABITAT AND HABITS** Found in primary and secondary, semi-deciduous monsoon, dipterocarp and tropical forests in north and north-east Borneo. The main diet comprises fruits, leaves, flowers and insects.

East Bornean Grey Gibbon

Kloss's Gibbon ■ *Hylobates klossii* HB 44–63cm, T absent

DESCRIPTION Both males and females have short black hair. Females are slightly larger in size than males. The species is broad chested and long limbed, and the hair on its head is flat. It is also known as the Dwarf Gibbon because it is the smallest gibbon species. **DISTRIBUTION** Indonesia (endemic to the four Mentawai Islands of Siberut, Sipora, North Pagai and South Pagai, off the west coast of Sumatra). **HABITAT AND HABITS** Found in tropical and monsoon rainforests. Diurnal and arboreal. The main diet comprises fruits, insects and leaves. Monogamous, living in small family groups of parents and offspring, with both parents caring for the young. **Notes** Both sexes sing in same sex choruses (males before dawn, females in mid-morning), where their solo songs overlap but are not synchronized with the songs of neighbouring groups.

Javan Gibbon ■ *Hylobates moloch* HB 45–64cm, T absent

DESCRIPTION Long, silver-grey fur (for this reason, it is also known as the Silvery Gibbon), with a slightly darker grey cap. Some individuals also have dark chests. Infants are a lighter grey than adults. **DISTRIBUTION** Indonesia (central and western Java). **HABITAT AND HABITS** Found in lowland to lower montane rainforests. Mainly eats fruits, leaves and flowers, and generally favours taller trees for foraging. Unlike other gibbons, it does not duet; the female is the main vocalist, while the male sings only occasionally. **Notes** Uniquely among gibbons, most of the singing is done by the female, with males singing occasionally in pre-dawn choruses.

Hainan Gibbon ■ *Nomascus hainanus* HB 48–54cm, T absent

DESCRIPTION Males have a distinct crest and are almost completely black, but females are beige, yellowish or orange with a dark, crestless cap. **DISTRIBUTION** China (Hainan Island). **HABITAT AND HABITS** Lives in tropical evergreen rainforest. Mainly feeds on fleshy fruits, but also on small animals and young leaves. **Notes** The world's rarest ape, and one of the most Critically Endangered mammals. Only one population of 23–25 individuals survives in a small area inside the Bawangling National Nature Reserve on the island of Hainan, off the west coast of China. Conservation efforts are ongoing, but the Bawangling population has shown extremely limited growth over the last three decades. None of this species is known to be in captivity.

Eastern Crested Gibbon ■ *Nomascus nasutus* HB 47–50cm, T absent

DESCRIPTION Males are completely black except for some sparse white hairs at the corner of the mouth. There is a prominent crest on the crown. Females have dark brown fur on the chest, sometimes reaching the stomach, and a lighter back. **DISTRIBUTION** China (south-west) and Vietnam (north-east). **HABITAT AND HABITS** Lives in lower montane and limestone forests. The single known population is restricted to limestone forest on karst outcrops. Its main diet is fruits. **Notes** Numbering just 50 surviving individuals, this Critically Endangered species is only known to exist in one area on the China–Vietnam border. It is mainly threatened by hunting, and habitat loss and degradation from fuelwood harvesting and livestock grazing. Conservation efforts are ongoing, with habitat restoration being a priority.

male

female

Western Black Crested Gibbon ■ *Nomascus concolor* HB 43–54cm, T absent

DESCRIPTION Males are almost entirely black, but sometimes have buff or white cheeks, whereas females are a yellow, orange, buff or golden colour with a black crown and dark patch on the ventral region. The species gets its common name from the crest-like tuft of fur on its head. **DISTRIBUTION** China (south Yunnan province), Laos (north-west) and Vietnam (north). **HABITAT AND HABITS** Evergreen and semi-evergreen forests. Diet comprises mainly fruits, leaves and insects. Lives in small family groups and is highly arboreal, like other gibbons. Utters loud calls in the morning to mark territories, with females singing in rising notes and ending with twitters, while males grunt, boom and whistle. **Notes** Highly threatened by hunting and habitat loss and fragmentation, this species has suffered a decline of more than 80 per cent in the last 45 years. It has also lost 75 per cent of its natural habitat. Only 1,300–2,000 individuals remain.

Northern White-cheeked Crested Gibbon

■ *Nomascus leucogenys* HB 45–64cm, T absent

DESCRIPTION Males are mostly black with scattered silver hairs, and white cheek-tufts that end at the same level as the tops of the ears without touching the mouth corners. Females are buff to orange, with a dark crown-patch and white face-ring. **DISTRIBUTION** China (south – Yunnan province), Laos (north) and Vietnam (north-west). **HABITAT AND HABITS** Lives in tall primary and semi-evergreen forests. Spends the majority of time in the canopy, hardly ever descending to the forest floor. Mainly eats fruits, leaves and insects, and comprises small family groups with a male, female and 3–4 offspring. **Notes** Probably extirpated from China.

Southern White-cheeked Crested Gibbon

Juvenile

■ *Nomascus siki* HB 45–55cm, T absent

DESCRIPTION As in the case of the Northern White-cheeked Crested Gibbon (see p. 158), males are generally black with scattered silver hairs, but they have white cheek-patches extending to the lower edges of the ears, touching the mouth at the bottom. Females are creamy-orange with dark crown-patches and a white facial ring. The species is shorter haired than the Northern White-cheeked Gibbon. **DISTRIBUTION** Laos (central) and Vietnam (central). **HABITAT AND HABITS** Lives in tall evergreen forests as well as steep karst forests. Main diet comprises fruits, leaves and insects. **Notes** There are some taxonomic disputes, and this may not be a genuine species but rather a natural hybrid of the Northern White-cheeked Crested and Southern Yellow-cheeked Crested Gibbons (see pp. 154 and 160).

Northern Yellow-cheeked Crested Gibbon

■ *Nomascus annamensis* HB 44–46cm, T absent

DESCRIPTION Males are generally black and females are buff coloured, very similar to the Southern Yellow-cheeked Crested Gibbon (see p. 160), from which this gibbon was recently split as a full species. Adult males have a brownish tinge on the chest, and buff cheeks extending to less than halfway up the ears, with rounded upper margins. Adult females are pale to orange-yellow with dark crown streaks and a darker chest. Infants are born whitish-buff, transitioning to the male pelage, and females changing to the buffy pelage when mature. **DISTRIBUTION** Cambodia (north-east), Laos (south) and Vietnam (central). **HABITAT AND HABITS** Found in broadleaved evergreen and semi-evergreen forests, and probably more deciduous adjacent habitats, with limited field data suggesting that it mainly eats fruits, and also leaves, shoots and flowers.

Southern Yellow-cheeked Crested Gibbon

Nomascus gabriellae HB 45–55cm, T absent

DESCRIPTION Males are black with buff-gold cheeks and a little reddish hair near the sides of the chin, which is mostly black. Females are buff coloured. Cheek hairs on both sexes appear to be brushed sideways. **DISTRIBUTION** Cambodia (south-east), Laos (south) and Vietnam (south). **HABITAT AND HABITS** Lives in tall evergreen and semi-evergreen, bamboo and woodland forests. Mainly eats fruits and leaves. Typical family group size is 3–5 individuals. **Notes** Poaching for the meat and pet trades is a major threat to this species.

Siamang *Symphalangus syndactylus* HB 75–90cm, T absent

DESCRIPTION The largest of the gibbons, this species has a stocky build relative to other gibbon species. It has shaggy black hair, and a greyish lower face. Both males and females have a throat pouch that visibly inflates when calling. Males also have a long scrotal tuft that resembles a short tail. **DISTRIBUTION** Indonesia (Sumatra), Malaysia (Peninsular Malaysia) and Thailand (southernmost, in the Thai-Malay Peninsula). **HABITAT AND HABITS** Found in primary and secondary forests, from lowlands up to 1,500m, in small family groups of a pair and offspring. Arboreal but moves less gregariously than other gibbons, instead brachiating more gracefully through the trees. Siamangs have a loud, booming call, followed by whooping, and males end their calls with a loud yell. **Notes** This species lives sympatrically with other gibbons, that is with the White-handed or Agile Gibbons (see pp. 142 and 143) in Sumatra and Peninsular Malaysia.

HOMINIDAE (ORANGUTANS)

Asia's only great apes are the orangutans – the only reddish-coloured apes and the largest tree-dwelling primates on Earth. There are two orangutan species, confined to only two range countries; the Sumatran Orangutan *Pongo abelli* and the Bornean Orangutan *P. pygmaeus*.

Orangutans are found in primary lowland swamp to montane forests, generally preferring altitudes below 500m. Both species survive on a diet of fruits, leaves and small animal prey. While both depend on high-quality primary forests, the Bornean Orangutan seems to be better able to tolerate some degree of habitat disturbance. In Sumatra, Indonesia, densities plummet by up to 60 per cent with even selective logging.

Orangutans are highly sexually dimorphic. Males are solitary, while females remain with their offspring. Orangutans have the longest childhood of any non-human primate. Offspring remain with the mother for up to eight years, even after she has another infant. Orangutans travel through the trees using a measured clambering technique known as quadrumanual climbing. They sleep in nests that they build out of broken twigs and branches woven together high in the trees.

The name orangutan means 'man of the forest' in Malay and Indonesian.

Bornean Orangutan

■ *Pongo pygmaeus* HB 780–970cm, T absent

DESCRIPTION Large bodied and generally heavier set than the Sumatran Orangutan (see p. 165), and has darker reddish-brown hair that varies from deep orange to maroon-brown. It also has a broader face with less facial hair than the Sumatran Orangutan. **DISTRIBUTION** Endemic to Borneo. Indonesia (south-west, central, east and north-west Kalimantan) and Malaysia (Sabah and north and south Sarawak). Not known to be resident in Brunei. **HABITS AND HABITAT** Lives mainly in lowland rainforests, and swamp and mountain forests. Peat swamps and forests that are prone to flooding produce larger and more regular fruit crops, and therefore hold the highest densities of orangutans. **Notes** There are three subspecies on Borneo. The north-west subspecies *P. p. pygmaeus*, with approximately 1,500 remaining individuals, is the most threatened mainly because of severe loss and fragmentation of its habitat, and hunting.

Bornean Orangutan

Sumatran Orangutan ■ *Pongo abelli* HB 780–970cm, T absent

DESCRIPTION Large-bodied ape with a reddish-brown shaggy coat, long limbs and no tail. It has bare, dark facial skin, though juveniles have pale pinkish facial skin. Males are much larger than females. Though generally very similar in appearance to the Bornean Orangutan (see p. 163), Sumatran Orangutans are thinner, and have a lighter red coat, longer hair and longer faces. **DISTRIBUTION** Indonesia (north-west Sumatra). **HABITS AND HABITAT** Almost completely arboreal – only adult males descend to the ground and do so very rarely. In the wild, it is estimated that males can live to up to 58 years of age, and females to 53. **Notes** Orangutans in Suaq Belimbing, Sumatra, are the only ones known to regularly make and use tools from sticks to extract termites and honey from hard-to-reach spots inside tree trunks. Though this kind of tool use has yet to be observed in Borneo, all orangutans use some tools to an extent, such as large leaves as umbrellas.

Sumatran Orangutan

References and Further Reading

While we obtained information for this book from a number of sources, including scientific papers, books and other sources, two that were the most useful were the Asian Primates Journal and Handbook of the Mammals of the World. Vol. 3. Primates, as listed below. A few other books that we found exceptionally useful included:

Francis, C. (2008). *A Guide to the Mammals of South-East Asia*. Princeton University Press, Princeton, New Jersey, and Oxford, United Kingdom.

Mittermeier, R.A., Rylands, A.B. and Wilson, D.E. eds (2013). *Handbook of the Mammals of the World. Vol. 3. Primates*. Lynx Editions, Barcelona.

Nadler, T., Momberg, F., Nguyen Xuan Dang and Lormee N. (2003): *Vietnam Primate Conservation Status Review 2002. Part 2: Leaf Monkeys*. Fauna & Flora International-Vietnam Program and Frankfurt Zoological Society, Hanoi.

Roos, C., Boonratana, R., Supriatna, J., Fellowes, J.R, Groves, C.P., Nash, S.D., Rylands, A.B. and Mittermeier, R.A. (2014). An updated taxonomy and conservation status review of Asian primates. Asian Primates Journal. Vol. 4, No. 1.

Rowe, N. (1996). The pictorial guide to the living primates. East Hampton, N.Y: Pogonias Press.

The Primate Specialist Group

The IUCN SSC Primate Specialist Groups is a network of scientists and conservationists working to save primates. Their website and related links and materials, are an excellent source of information. http://www.primate-sg.org/

Acknowledgements

This book would not have been possible without the input, support and advice from individuals working on various primate conservation efforts throughout Asia and the very generous contributions of photographs from these same individuals, as well as many other photographers and naturalists. All photographers are credited on page 4. However, we would like to especially thank Andie Ang, James Eaton, Tilo Nadler, Anna Nekaris, Ramesh 'Zimbo' Boonratana, Gabriella Fredriksson and members of the IUCN SSC Primate Specialist Group for helping us to track down hard-to-come-by images. Our deepest gratitude goes to Vincent Nijman and Will Duckworth for their advice and guidance, from beginning to the end. We would also like to thank John Beaufoy and Rosemary Wilkinson for pulling this whole project together.And to the conservationists and researchers working to better understand and protect Asia's primates – thank you.

PHOTOGRAPHS

Despite the best endeavours of the authors and publishers, we were unable to track down photographs of the Black-and-White Langur, Hose's Langur or Abbott's Grey Gibbon. The publishers would be delighted to hear from anyone who might be able to provide photographs of these species for inclusion in a subsequent edition of the book.

Checklist of the Primates of Southeast Asia

For each species, an 'x' indicates presence in a particular country; where this is not confirmed, there is a question mark. Country abbreviations are as follows:

BD Bangladesh
BN Brunei
BT Bhutan
CN China
ID Indonesia
IN India
JP Japan
KH Cambodia
LA Lao PDR
LK Sri Lanka
MM Myanmar
MY Malaysia
NP Nepal
PH Philippines
PK Pakistan
SG Singapore
TH Thailand
TL Timor-Leste
VN Vietnam

English name	Scientific name	BD	BT	BN	KH	CN	IN
Tarsiidae (Tarsiers)							
Spectral Tarsier	*Tarsius tarsier*						
Makassar Tarsier	*Tarsius fuscus*						
Dian's Tarsier	*Tarsius dentatus*						
Peleng Tarsier	*Tarsius pelengensis*						
Sangihe Tarsier	*Tarsius sangirensis*						
Siau Island Tarsier	*Tarsius tumpara*						
Pygmy Tarsier	*Tarsius pumilus*						
Lariang Tarsier	*Tarsius lariang*						
Wallace's Tarsier	*Tarsius wallacei*						
Philippine Tarsier	*Carlito syrichta*						
Western Tarsier	*Cephalopachus bancanus*						
Lorisidae (Lorises)							
Red Slender Loris	*Loris tardigradus*						
Grey Slender Loris	*Loris lydekkerianus*						
Bengal Slow Loris	*Nycticebus bengalensis*	X			X	X	X
Sunda Slow Loris	*Nycticebus coucang*						
Javan Slow Loris	*Nycticebus javanicus*						
Philippine Slow Loris	*Nycticebus menagensis*			X			
Bangka Slow Loris	*Nycticebus bancanus*						
Bornean Slow Loris	*Nycticebus borneanus*						
Kayan Slow Loris	*Nycticebus kayan*						
Pygmy Slow Loris	*Nycticebus pygmaeus*				X	?	
Cercopithecidae (Colobines)							
Thomas's Langur	*Presbytis thomasi*						
Black-crested Sumatran Langur	*Presbytis melalophos*						
Black Sumatran Langur	*Presbytis sumatrana*						
Black-and-White Langur	*Presbytis bicolor*						
Mitred Langur	*Presbytis mitrata*						
Javan Grizzled Langur	*Presbytis comata*						
Pagai Langur	*Presbytis potenziani*						
Siberut Langur	*Presbytis siberu*						
Common Banded Langur	*Presbytis femoralis*						
White-thighed Langur	*Presbytis siamensis*						
Natuna Islands Langur	*Presbytis natunae*						
Bornean Banded Langur	*Presbytis chrysomelas*			X			
Maroon Langur	*Presbytis rubicunda*			?			
Hose's Langur	*Presbytis hosei*			X			
Miller's Grizzled Langur	*Presbytis canicrus*						
Sabah Grizzled Langur	*Presbytis sabana*						
White-fronted Langur	*Presbytis frontata*						
Gee's Golden Langur	*Trachypithecus geei*		X				X
Capped Langur	*Trachypithecus pileatus*	X	X				X

	ID	JP	LA	MY	MM	NP	PK	PH	SG	LK	TH	TL	VN
	X												
	X												
	X												
	X												
	X												
	X												
	X												
	X												
	X												
								X					
	X			X									
										X			
										X			
			X	?	X						X		X
	X			X					X		X		
	X												
	X			X					X				
	X												
	X												
	X			X									
			X										X
	X												
	X												
	X												
	X												
	X												
	X												
	X												
	X												
	X			X	X				X		X		
	X			X							X		
	X												
	X			X									
	X			X									
	X			X									
	X												
				X									
	X			X									
					X								

English name	Scientific name	BD	BT	BN	KH	CN	IN
Shortridge's Langur	*Trachypithecus shortridgei*					X	
Eastern Ebony Langur	*Trachypithecus auratus*						
Ebony Langur	*Trachypithecus mauritius*						
Sundaic Silvered Langur	*Trachypithecus cristatus*			X			
Selangor Silvered Langur	*Trachypithecus langorensis*						
Indochinese Silvered Langur	*Trachypithecus germaini*				X		
Annamese Langur	*Trachypithecus margarita*				X		
Dusky Langur	*Trachypithecus obscurus*						
Phayre's Langur	*Trachypithecus phayrei*	X				X	X
Indochinese Grey Langur	*Trachypithecus crepusculus*					X	
Tenasserim Langur	*Trachypithecus barbei*						
Francois' Langur	*Trachypithecus francoisi*					X	
Cat Ba Langur	*Trachypithecus liocephalus*						
White-headed Langur	*Trachypithecus cocephalus*					X	
Delacour's Langur	*Trachypithecus delacouri*						
Lao Langur	*Trachypithecus laotum*						
Hatinh Langur	*Trachypithecus hatinhensis*						
Black Langur	*Trachypithecus ebenus*						
Northern Plains Grey Langur	*Semnopithecus entellus*						X
Kashmir Grey Langur	*Semnopithecus ajax*						X
Terai Grey Langur	*Semnopithecus hector*		?				X
Central Himalayan Sacred Langur	*Semnopithecus schistaceus*		X			X	X
Black-footed Langur	*Semnopithecus hypoleucos*						X
Tufted Grey Langur	*Semnopithecus priam*						X
Purple-faced Langur	*Semnopithecus vetulus*						
Nilgiri Langur	*Semnopithecus johnii*						X
Tonkin Snub-nosed Monkey	*Rhinopithecus avunculus*					?	
Black Snub-nosed Monkey	*Rhinopithecus bieti*					X	
Grey Snub-nosed Monkey	*Rhinopithecus brelichi*					X	
Golden Snub-nosed Monkey	*Rhinopithecus roxellana*					X	
Myanmar Snub-nosed Monkey	*Rhinopithecus strykeri*					X	
Grey-shanked Douc	*Pygathrix cinerea*				X		
Red-shanked Douc	*Pygathrix nemaeus*				X		
Black-shanked Douc	*Pygathrix nigripes*				X		
Proboscis Monkey	*Nasalis larvatus*			X			
Pig-tailed Langur	*Simias concolor*						
Cercopithecidae (Macaques)							
Lion-tailed Macaque	*Macaca silenus*						X
Northern Pig-tailed Macaque	*Macaca leonina*	X			X	X	X
Southern Pig-tailed Macaque	*Macaca nemestrina*			X			
Siberut Macaque	*Macaca siberu*						
Pagai Island Macaque	*Macaca pagensis*						

ID	JP	LA	MY	MM	NP	PK	PH	SG	LK	TH	TL	VN
				X								
X												
X												
X			X									
			X									
		X		X						X		X
		X										X
			X	X						X		
				X								
		X		X						X		X
				X						X		
												X
												X
												X
		X										
		X										X
		X										X
					X							
					X	X						
									X			
									X			
												X
				X								
		?										X
		X										X
		?										X
X			X									
X												
		X		X						X		X
X			X							X		
X												
X												

English name	Scientific name	BD	BT	BN	KH	CN	IN
Black-crested Macaque	*Macaca nigra*						
Gorontalo Macaque	*Macaca nigrescens*						
Tonkean Macaque	*Macaca tonkeana*						
Booted Macaque	*Macaca ochreata*						
Heck's Macaque	*Macaca hecki*						
Moor Macaque	*Macaca maura*						
Toque Macaque	*Macaca sinica*						
Bonnet Macaque	*Macaca radiata*						X
Assamese Macaque	*Macaca assamensis*	X	X			X	X
Tibetan Macaque	*Macaca thibetana*					X	?
Arunachal Macaque	*Macaca munzala*		?			?	X
White-cheeked Macaque	*Macaca leucogenys*						
Stump-tailed Macaque	*Macaca arctoides*	X			X	X	X
Long-tailed Macaque	*Macaca fascicularis*	X		X	X		X
Rhesus Macaque	*Macaca mulatta*	X	X			X	X
Japanese Macaque	*Macaca fuscata*						
Formosan Rock Macaque	*Macaca cyclopis*					X	
Hylobatidae (Gibbons)							
Western Hoolock	*Hoolock hoolock*	X					X
Gaoligong Hoolock	*Hoolock tianxing*						
Eastern Hoolock	*Hoolock leuconedys*					X	X
Pileated Gibbon	*Hylobates pileatus*				X		
White-handed Gibbon	*Hylobates lar*					X	
Agile Gibbon	*Hylobates agilis*						
Bornean White-bearded Gibbon	*Hylobates albibarbis*						
Mueller's Bornean Gibbon	*Hylobates muelleri*						
Abbott's Grey Gibbon	*Hylobates abbotti*						
East Bornean Grey Gibbon	*Hylobates funereus*						
Kloss's Gibbon	*Hylobates klossii*						
Javan Gibbon	*Hylobates moloch*						
Hainan Gibbon	*Nomascus hainanus*					X	
Eastern Crested Gibbon	*Nomascus nasutus*					X	
Western Black Crested Gibbon	*Nomascus concolor*					X	
Northern White-cheeked Crested Gibbon	*Nomascus leucogenys*					X	
Southern White-cheeked Crested Gibbon	*Nomascus siki*						
Northern Yellow-cheeked Crested Gibbon	*Nomascus annamensis*				X		
Southern Yellow-cheeked Crested Gibbon	*Nomascus gabriellae*				X		
Siamang	*Symphalangus syndactylus*						
Hominidae (Orangutans)							
Bornean Orangutan	*Pongo pygmaeus*						
Sumatran Orangutan	*Pongo abelli*						

ID	JP	LA	MY	MM	NP	PK	PH	SG	LK	TH	TL	VN
X												
X												
X												
X												
X												
X												
									X			
		X		X	X	X				X		X
		X	X	X						X		X
X		X	X	X			X	X		X	X	
		X		X	X	X				X		X
	X											
				X								
				X								
		X								X		
X		X	X	X						X		
X			X							X		
X												
X												
X			X									
X			X									
X												
X												
												X
		X										X
		X										X
		X										X
		X										X
		X										X
X			X							X		
x			x									
x												